Generis

PUBLISHING

Introduction to Electromagnetism According to Maxwell

(Electromagnetic mechanics)

André Michaud

CIP a Camerei Naționale a Cărții

Michaud, André.

Introduction to Electromagnetism According to Maxwell : (Electromagnetic mechanics)/André Michaud. – Chișinău : Generis Publishing, 2020 (Print on demand). – 265 p. : fig., tab.

Tit. orig.: Introduction à l'électromagnétisme selon Maxwell. - Referințe bibliogr.: p. 255-264 (100 tit.).

ISBN 978-9975-3238-3-3.

537.8

M 65

Cover image: www.pixabay.com

Generis Publishing
Online orders: www.generis-publishing.com
Orders by email: info@generis-publishing.com

"Things happen in this world when

someone makes them happen"

Table of Contents

Foreword

For the first mechanical explanation of electromagnetic photons emission and absorption by electrons to currently make sense in the physics community, the explanation can be made at this point in time only from four unfamiliar aspects of electromagnetism, two of which are very recent developments that are unfamiliar for this very reason, which are the trispatial geometry that was proposed in 2000 and Paul Marmet's derivation that was published only 3 years later, both of which must be correlated with Louis de Broglie's hypothesis about localized photon's possible inner electromagnetic structure and Maxwell's initial conclusion that both electric and magnetic fields have to induce each other for the existence of electromagnetic energy to be correctly described.

Unfortunately, both de Broglie's hypothesis and Maxwell's initial interpretation, although formally available in the literature, are themselves unfamiliar to most in the current physics community, which is why the sequence of arguments presented in Chapter 1 of this work is arranged in such a way as to progressively relate these four unfamiliar aspects with the main familiar conclusions previously drawn about elementary particles, to make more obvious how these four unfamiliar aspects harmonize with observation, and can consequently be used as a solid foundation to ultimately explain photon emission and absorption.

This unfamiliarity with the conclusions of Maxwell and de Broglie is mainly due to the dominance for the past century of the Copenhagen interpretation, a dominance that eventually became so absolute in the orthodox physics community, that many of the major seminal papers that were published by Max Planck, Albert Einstein and Louis de Broglie, among other major contributors to the advancement of knowledge in physics, who opposed this interpretation, are no longer referred to, and to this day, have not even been translated to English to be made available to the global physics community. Nowhere is the nefarious influence of the Copenhagen interpretation on the physics community better put in perspective than in an analysis published initially in German by Franco Selleri, subsequently translated to French and Spanish, under the title of *"Die Debatte um die Quantentheorie"* (*"The Debate about Quantum Theory"*) [1].

This translation issue is currently in process of being addressed by organizations such as the *Minkowski Institute Press* founded by Vesselin Petkov, dedicated to making available in English many of these ground breading papers. Among the impressive list of such untranslated papers, my friend Fritz Lewertoff, who contributed in 2012 the first ever translation to English of Herman Minkowski's *"Das Relativitätsprinzip"* (*"The Relativity Principle"*) [2], made me aware of two other major papers in this list, whose earlier translation could possibly have allowed progress to resume much sooner in fundamental physics, and that are now in process of being translated.

The first one is the text of a lecture given by Max Planck on November 12, 1930, titled *"Positivismus und reale Aussenwelt"* [3] (*"Positivism and the real outside world"*), in which he exposes the manner in which skepticism had been gaining ground in fundamental physics to the point of casting doubts on logical reasoning itself, and how such an attitude, that he had just witnessed being promoted 3 years before during the 1927 Solvey congress, was likely to mislead the community into the absence of progress that we have been observing for decades now in fundamental physics research. This damaging philosophy, that was actively promoted by Bohr, Heisenberg and Sommerfeld, became eventually known as the *"Copenhagen interpretation"*, and, to the chagrin of all in the community who believe in the benefits of rationality, has become the dominant philosophy in the orthodox fundamental physics community for the past 90 years.

The most striking statement in Planck's lecture is a remark that certainly was meant as a warning about the dangers of this skepticism with regard to logical reasoning that was gaining more and more ground at that time in the fundamental physics community, according to which we will never be able to understand reality at the fundamental level any more clearly than the vague outlines allowed by Heisenberg's statistical description method, which is an axiomatic dogma directly contradicted by the current state of our understanding of the subatomic level from the electromagnetic perspective:

> *"Ein Menschenkind, das seine eigene Zukunft als durch das Schicksal zwangsläufig vorherbestimmt ansieht, oder ein Volk, das den Prophezeiungen seines naturgesetzlich festgelegten Unterganges Glauben schenkt, bekundet damit in Wirklichkeit nur, daß es den rechten Willen zum Aufstieg nicht aufzubringen vermag."* ([3], p. 34).

Translation:

"A human being who sees his own future as inevitably predetermined by fate, or a people who believes the prophecies of its downfall determined by natural law, in reality only shows that it cannot muster the right will to ascend."

Planck's concern about this loss of confidence in logical reasoning that seemed to become the orthodox belief in the fundamental physics community soon proved to have been justified, and already in 1953 Schrödinger bluntly denounced it in a work that still has not been translated to English to be made available to the international community ([4], p. 16). See quote of this denunciation in Section 2.1.

Planck's analysis clearly highlights the limited range of possibilities for progress offered by the statistical approach that was gaining ground in the physics research community compared to those offered by the dynamic approach, in the clear identification of the laws of nature.

The second text is an incredibly important paper from Albert Einstein dating back to 1910 [5], and that practically nobody has read nor referred to for the past century, for the simple reason that the only existing version of this text is a translation to French of the lost German original, titled *"Le Principe de relativité et ses conséquences dans la physique moderne"* (*"The Principle of Relativity and its Consequences in Modern Physics"*).

The importance of this paper lies in the fact that it reveals that as early as 1910, Einstein already was aware of the 1:1 identity relation that exists between the electrodynamic force related to the acceleration of the electron charge e when subjected to an E-field, and the gravitational force related to the acceleration of mass m of the same electron, as established by Newton for macroscopic masses, which he summarized with Equation (2) on page 143 of this paper:

"On peut, par exemple, obtenir de cette façon les équations du mouvement d'un point matériel de masse m portant une charge électrique e (par exemple un électron) et soumis à l'action d'un champ électromagnétique. On connaît, en effet, les équations du mouvement d'un point matériel à l'instant où sa vitesse est nulle. D'après les équations de Newton et la définition de l'intensité du champ électrique, on a:"

Translation:

"We can, for example, obtain in this way the equations of motion of a material point of mass m carrying an electric charge e (for example an electron) and subjected to the action of an electromagnetic field. We know, in fact, the equations of motion of a material point at the moment when its velocity is zero. According to Newton's equations and the definition of the electric field strength, we have:"

$$(2) \qquad m\frac{\mathrm{d}^2 x}{\mathrm{d}t^2} = e\mathbf{E}_x \qquad\qquad \text{([5], p. 143)}$$

This correct understanding on his part of the relation between the invariant rest mass and the invariant charge of the electron certainly explains his persisting intuition that gravitation had to be related to electromagnetism, as we will further analyze in Section 1.7.1. It is well known that towards the end of his life, he had became adamant that gravitation had to be related to electromagnetism, and was openly advocating that this avenue should be investigated, even if this meant that his brainchildren theories of Special Relativity (SR) and General Relativity (GR) had to be abandoned as physically inapplicable, that is, even if his theories ultimately turned out to only be *"a castle in the air"*, as he wrote in 1954 [6].

In fact, the development of these *relativity* theories at the beginning of the 20th century came about due to an alleged impossibility of demonstrating absolute motion in the universe, giving precedence to the concept *of relative motion* as opposed to *absolute motion*, that was brought to general attention by mathematician Henri Poincaré in a short note widely distributed by the French *Académie des Sciences*, in June of 1905. This issue will be addressed in Section 3.4, and Subsections 3.5.1 and 3.17.1.

Unfortunately, when Einstein formulated this recommendation that more attention should be given to electromagnetism a few years before he passed away in 1955, the Copenhagen interpretation had already conquered the whole fundamental physics research domain, as confirmed by Schrödinger's denunciation in 1953 (See Section 2.1), and the whole orthodox community apparently purposefully immediately rejected his recommendation without a second look, as reported in 1995 by Archibald Wheeler, a major Copenhagen interpretation opinion leader:

"A distinguished physicist even published in his very last years' works, the main point of which is to claim that gravitation follows

the pattern of electromagnetism. This thesis, we cannot accept, and the community of physics, quite rightly, does not accept."

Archibald Wheeler, 1995. ([7], p. 391)

The unfortunate outcome of this outright rejection was a 40 years hiatus before this investigation could be re-initiated in the late 1990's, right after this author became aware of this comment by Wheeler in the work that he co-authored and published in 1995 with Ignazio Ciufolini [7]. This apparently incomprehensible refusal to proceed with fundamental research in such an important direction will be analyzed in Section 1.7.2.

The project that the present work is part of is meant to repair the damage caused by this rejection, by exploring and analyzing the subatomic magnitude level of physical reality from the long established experimental foundations of electromagnetism, by means of an expansion to Maxwell's 3D vectorial space. Among the various aspects of the subatomic level that will be analyzed, Sections 1.26 and 1.27 cover what the study of electromagnetism leads to with regard to gravitation, apparently confirming that Einstein's conclusion that gravitation follows the pattern of electromagnetism may well have been right.

Most of the freely available previously published papers in this project, that refocus the conclusions drawn about the various observed phenomena at the subatomic level according to this new perspective, have been regrouped in a monograph published separately [8]. The three remaining articles that were subsequently published, also in open access, including the final synthesis of the project, are now being regrouped in the present work.

Chapter 1 reproduces the content of the article cited as Reference [9] Titled *"Electromagnetism according to Maxwell's Initial Interpretation"* formally published in January of 2020 and that constitutes the final synthesis of this project. The required sequence of arguments is organized in this chapter so as to progressively connect all four unfamiliar aspects initially mentioned with the main familiar conclusions previously drawn about elementary particles, to make more obvious to what extent these unfamiliar aspects harmonize with observation, and can consequently be used as a solid foundation to ultimately explain photon emission and absorption.

Chapter 2 reproduces the content of the article cited as Reference [10] titled *"The Hydrogen Atom Fundamental Resonance States"*, formally published in April 2018. It retraces the origins of Quantum Mechanics and refocuses its understanding according to the conclusions of its initial originators, who were Louis de Broglie and Erwin Schrödinger, to finally explain, in context of the previously mentioned expanded space geometry, why electrons cannot crash onto atomic nuclei in Nature, but are rather captured in various stable stationary action orbitals at some distances from these nuclei.

Finally, Chapter 3 reproduces, with a few complementary Subsections, the content of the article cited as Reference [11] titled *"Gravitation, Quantum Mechanics and the Least Action Electromagnetic Equilibrium States"* formally published in November 2017. It provides a simplified overview of the states and processes described in the series of articles that were regrouped in the monograph titled *"Electromagnetic Mechanics of Elementary Particles"*, that was published separately in 2017 [8]. So that the present introduction to electromagnetism can serve as an index into both the complete series of freely available papers, and also into the related monograph, all references to the separate papers will also refer the specific chapters that integrate them into the monograph, for readers who prefer to use the integrated monograph.

A very positive development regarding the latter article is that it was chosen as one of the chapters of the eBook titled *"Prime Archives in Space Research"*, to be republished by the *Vide Leaf Prime Archives*, whose aim is to promote scientific research in the world by making research results considered state-of-the-art available to young researchers to facilitate their application in their research practices. This choice can only hasten the re-familiarization of the community with Maxwell's initial interpretation and a better understanding of physical reality that it seems to favor. This republication is cited as Reference [12].

A certain amount of overlap of the descriptions will be observed between all three chapters, but since each chapter reproduces the actual content of a separately published paper, it was chosen not to reduce these overlaps so as not to interfere with the equations numbering sequences, and most importantly, not to interfere with the specific lines of reasoning that each paper was meant to emphasize. This allows all three chapters to remain independent of each other so they can be read in any order without prejudice.

1. Electromagnetism according to Maxwell's Initial Interpretation

1.1 Introduction

It is well established that classical electrodynamics, quantum electrodynamics (QED) as well as Quantum Field Theory (QFT) are grounded on Maxwell's wave theory and on his equations, but it is much less well understood that they are not grounded on his initial interpretation of the relation between the E and B fields, but are rather grounded on Ludwig Lorenz's interpretation of this relation, with which Maxwell disagreed.

Maxwell considered that both fields had to mutually induce each other cyclically for the velocity of light to be maintained while Lorenz considered that both fields had to synchronously peak at maximum at the same time for this velocity to be maintained, both interpretations being equally consistent with the equations. Two recent breakthroughs however now allow confirming that Maxwell's interpretation was correct, at least with respect to the subatomic level, because, contrary to the Lorenz interpretation, it allows to seamlessly reconcile Maxwell's electromagnetic wave theory, so successfully applied at our macroscopic level, with the electromagnetic characteristics that apply at the subatomic level to localized electromagnetic photons and to all localized charged and massive elementary electromagnetic particles of which all atoms are made, and finally allows establishing a clear mechanics of electromagnetic photon emission and absorption by electrons during their interaction at the atomic level.

In 1845, Michael Faraday observed that by placing a glass plate between the poles of an electromagnet, the magnetic field caused the polarization plane of the light passing through the plate to rotate. He immediately informed his friend James Clerk Maxwell of this major discovery that demonstrated for the first time the direct relation between the magnetic field and light [13].

It is therefore this specific experiment by Faraday which is at the origin of the integrated electromagnetic theory then developed by Maxwell, because, having already observed that second derivatives of the previously established equations for the electric field and the magnetic field revealed that electric energy and magnetic energy were separately associated with the speed of light ([14], [8] Chapter 13), Maxwell concluded that light had to be electromagnetic in nature and then made the fundamental discovery that electromagnetic energy involves a three-way orthogonal relationship between its three fundamental aspects, that is, its electric and magnetic aspects, perceived as being perpendicular to each other and simultaneously inducing each other in a cyclic transverse stationary oscillating motion, with respect to the direction of motion of this energy in space (see **Figure 1.1**), that is, a three-way orthogonal relationship corresponding to the familiar vector cross product of the E and B fields (See **Figure 1.3-a**), resulting in a third motion vector structurally perpendicular to the first two ([15], [8] Chapter 6).

The following fact may come as a surprise to many, but this solution discovered by Maxwell, who is also well known for having derived the speed of light from the relation that he established between the two fundamental constants of vacuum ε_0 and μ_0 ([14], [8] Chapter 13), is not the only working solution that was discovered to relate both E and B fields to the speed of light.

Briefly summarized, mathematician Ludwig Lorenz established independently from Maxwell that if both E and B fields representations of free moving electromagnetic energy are mathematically made to peak to maximum synchronously at the same time, this also allows explaining the speed of light in vacuum of electromagnetic waves just as well as when both fields are 180° out of phase as in Maxwell's solution.

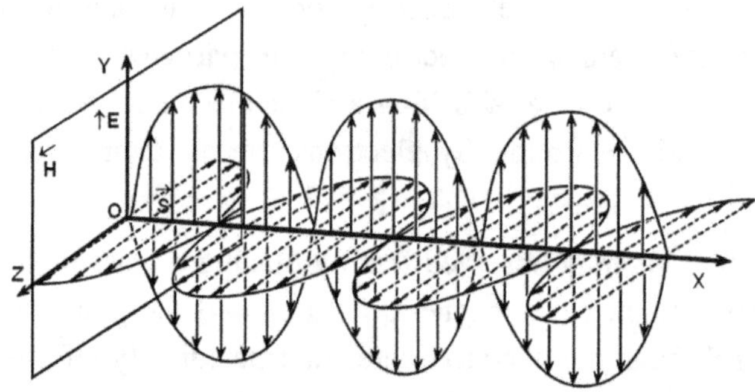

Figure 1.1: Mutually inducing 180° out of phase bipolar representation of E and B fields of Maxwell's interpretation.

But the *Lorenz gauge* is a generalizing concept, that regroups both E and B aspects of fundamental energy into a *single* electromagnetic field that distracts from immediate attention the different vectorial orientations of both aspects, particularly the fact that the energy dipole represented by E becomes spacewise oriented and distributed while the energy dipole represented by B becomes timewise oriented and distributed as they cyclically mutually induce each other transversely to the vectorial direction of motion of the oscillating energy in vacuum, as can be concluded from Maxwell's interpretation.

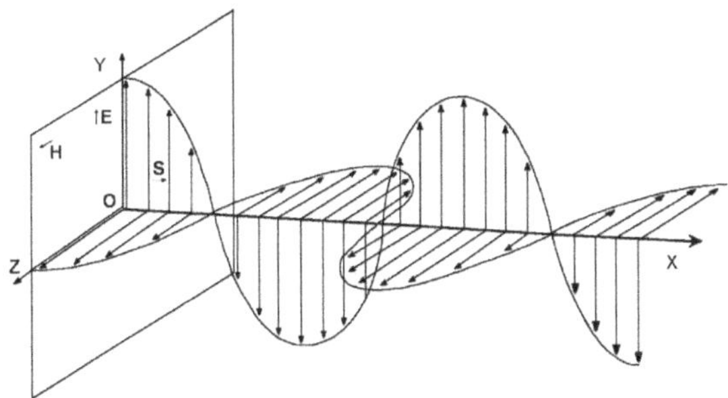

Figure 1.2: Standard simultaneously in phase peaking E and B fields monopolar representation of Lorenz's interpretation.

The representation of **Figure 1.2**, which is found in all textbooks on electromagnetism, although agreeing with Maxwell's wave theory describing electromagnetic energy as a pulse propagating in an underlying aether, and which is also in agreement with his equations, is however generally and incorrectly assumed as also being Maxwell's conclusion.

In fact, Maxwell disagreed with this approach, because the *gauge* concept developed by Lorenz had for consequence of treating both E and B fields as being a *single electromagnetic field* at the general level, which, at first glance, does not suggest any apparent internal structure, which easily obscures the fact that both fields are of separate and of equal importance in Maxwell's theory, with different and irreconcilable characteristics, in addition to mutually inducing each other, contrary to the Lorenz solution, as put into perspective in Reference ([15], [8] Chapter 6).

The fact that this second solution was developed by Lorenz, however, is not well known in the scientific community because it is specifically associated only to the so-called *Lorenz gauge* defined by him, and this, only in high level specialized reference works on electromagnetism [16], because it lends itself more easily than Maxwell's representation to various mathematical generalization processes as applicable at our macroscopic level, but the true origin of the solution represented by **Figure 1.2** is not clearly explained in introductory textbooks and general reference works on physics [17] [18].

Consequently, unless they specialize in electromagnetism, most physicists are not directly informed that it was not Maxwell who developed this second approach, and that classical electrodynamics and quantum field theory (QFT), from which quantum electrodynamics (QED) emerged [19] [20] are in reality grounded on Lorenz's interpretation, because this fact is nowhere clearly highlighted in reference works on electrodynamics and QFT, which were of course developed by specialists in electromagnetism for whom this fact was obvious. So contrary to established facts, the outcome is a general impression in the community that Maxwell is also the author of this second solution and that electrodynamics and QFT are grounded strictly on Maxwell's theory.

The distinction to be made is important however, because de Broglie's hypothesis about the localized double-particle photon as applicable to the subatomic level, that emerges directly from Maxwell's solution, is consequently at odds with classical electrodynamics and QED, because the Lorenz approach obscures the fact that both E and B fields are of equal and separate importance. For example, the predominant role given to the electric charges in QED seems to leave no precise function to the magnetic aspect of electromagnetic energy in a possible mutual induction mechanics that would involve the two separate fields, contrary to Maxwell's interpretation. Even the fact that as formulated, QED cannot explain the mutual induction of both fields in LRC systems doesn't seem to attract attention to this issue.

1.2. Setting up the perspective according to relative magnitude levels

To put in correct perspective the description of the energy which is the very substance of which all localized elementary particles such as electromagnetic photons, electrons and positrons are made at the subatomic level, in a manner that would not conflict with the well established Maxwell continuous wave electromagnetic theory, which is so successfully applied at our macroscopic level from the Lorenz perspective, it must first be realized that all objects and processes that we can detect and measure in objective reality can be categorized as belonging to one of the following four orders of magnitude. In decreasing order, these orders of magnitude can be defined very generally as follows:

1- *Astronomical level*: Order of magnitude exceeding the dimensions of planet Earth.

2- *Macroscopic level*: Order of magnitude in which any object or process can be directly measured at the Earth's surface and its environment.

3- *Sub-microscopic* or *atomic level*: Order of magnitude of molecules and atoms.

4- *Subatomic level*: Order of magnitude of the elementary particles of which all atoms are made, as well as the electromagnetic energy of which their substance is made, that supports their motion, determines their inertia, and that can also circulate as free moving quanta at the speed of light when not directly associated with one of these elementary particles.

The first 3 levels are generally familiar to all, but the subatomic level is not. We can directly perceive and measure objects and processes in our environment at the macroscopic level, and we indirectly perceive and measure objects and processes of the two neighboring orders of magnitude with increasing precision as our instruments improve, but we have no means to observe the fourth level.

It may seem paradoxical to so firmly assert that electromagnetic energy can be directly defined as being quantized as localized electromagnetic photons at the subatomic level in full accordance with Maxwell's equations while remaining in complete harmony with his continuous electromagnetic waves theory which has been so successfully applied at our macroscopic level, which is an issue that has been the object of a continuous debated for the past hundred years.

It must be put in perspective here that we perceive however no paradox whatsoever with the fact that *we directly observe* that the image on a TV screen appears smoothly continuous as seen from a few meters, while being well aware that if we get close enough, *we also directly observe*, directly at our macroscopic level, that in physical reality, the image is physically generated by thousands of clearly separated rows of clearly separated very small pixels.

Interestingly, we find no paradox either in treating water as a fluid without any internal structure at our macroscopic level, while being well aware that at the submicroscopic level, it is made only of localized molecules, themselves made of localized atoms, that we know are themselves made at the subatomic level of localized elementary electrically charged electrons, plus nucleons, themselves made of localized electrically charged elementary particles, that all are individually massive and quantized, even if we cannot directly see these molecules at our macroscopic level as in the case of the TV screen.

The reason why we see no problem in perceiving and treating water as a fluid at the macroscopic level, even mathematically, even if we cannot directly observe the localized molecules of which its substance is made, as we can directly do with the individual pixels of the TV screen, is that we understand that what we perceive as the *fluidity* of water at our macroscopic level is in reality a *crowd effect* due to countless localized water molecules smoothly sliding against each other at the submicroscopic level. Moreover, our powerful modern instruments of electronic microscopy allow us to indirectly detect these individual molecules and the atoms of which they are made at the submicroscopic level.

In the case of electromagnetic energy, however, its granular nature at the subatomic level is far from being as obvious to perceive as in the case of the television screen, in which to approach the image by only a few meters, is sufficient, to go from the order of magnitude that lets us perceive it as an apparently uniformly fluid image to the slightly lower order of magnitude still at the macroscopic level that makes it possible to perceive the reality of its granular structure when directly observed at greater proximity; or in the case of water, whose granularity at the atomic level can be indirectly observed with our electron microscopes.

The case of water obviously requires an even greater jump in orders of magnitude towards the infinitely small scale between the perception of its fluidity at the macroscopic level and the perception of its submicroscopic granularity. To really become aware of the difference between these two orders of magnitude, it suffices to think that the atoms making up water molecules are as far away towards the extremely small submicroscopic level that the galaxies are far towards the infinitely large astronomical level with respect to our own terrestrial macroscopic level. But to perceive the subatomic granularity of electromagnetic energy, the jump from our macroscopic order of magnitude is larger yet; as far in fact, further down towards the infinitely small from the already far order of magnitude of the atomic scale than this atomic scale is far from our own macroscopic level.

To really conceptualize how far down from the atomic scale the granularity of electromagnetic energy actually is, let's consider that if the proton of a hydrogen atom, two of which are part of a water molecule, was enlarged to become as big as the sun, the electron which is stabilized in its least action orbital distance from the proton would then be as far away from this enlarged proton as the orbit of Neptune is from the Sun in the Solar system, meaning that the hydrogen atom would become as large as the entire Solar System, and that the electromagnetic photons that constitute the *granular level* of electromagnetic energy is of the same order of magnitude as the energy making up the rest mass of the electron and of the other massive elementary electrically charged electromagnetic particles that exist inside the structure of the proton and of the neutron.

The main problem that we are confronted with regarding this subatomic level of granularity of electromagnetic energy and of the energy constituting the rest masses of elementary particles of which atoms are made, is that there exists no instrument powerful enough that would allow observing even indirectly this subatomic level, unlike the deepest level at which it remains physically possible, which is the atomic order of magnitude, that allows indirectly verifying the granularity of water and of all the other material substances of our environment; in short, an indirectly verifiable granularity of all atoms of the periodic table, which is unavailable for the subatomic granularity level of electromagnetic energy.

The only physically verifiable telltales that we have of the permanent localization of elementary charged particles such as the electron and of electromagnetic energy quanta are the following:

1- We have easily reproducible experimental proof that electrons and electromagnetic photons systematically behave almost point-like during all scattering experiments (See Section 1.23 further on and Reference [21]).

2- We have easily reproducible proof that photons have longitudinal inertia as demonstrated by the Einstein photoelectric experiment, and that they have transverse inertia amounting to half their longitudinal inertia, as demonstrated by the deflexion angle of light by the Sun during numerous experiments carried out during solar eclipses ([15], [8] Chapter 6) [22].

3- We also have experimental proof since 1933 that electromagnetic photons of 1.022 MeV or more convert into electron-positron pairs when they graze massive particles [23] and that such pairs reconvert to electromagnetic photons when meeting again; which means that we have the experimental proof that the invariant rest mass of electrons and positrons is made of the same *electromagnetic energy substance* as electromagnetic photons. We also have experimental proof since 1997 that electromagnetic photons that exceed the 1.022 MeV energy threshold level can be destabilized into converting to electron-positron pairs by other electromagnetic photons, without any massive nuclei being close by [24].

4- We have easily reproducible experimental proof that free moving electrons have an invariant rest mass of 9.10938188E-31 kg and an invariant electric charge of 1.602176462E-19 C.

5- We have conclusive experimental evidence that electrons are elementary particles and that the protons and neutrons that constitute the nuclei of all atoms are not elementary particles, but are rather systems of elementary particles (see **Figures 1.5, 1.6** and **1.7**, and Reference [21]).

Since the subatomic level cannot be directly nor indirectly observed, we are therefore necessarily reduced in our exploration of this level to proceed by reverse engineering ([10], See also Section 2.19 for more on this issue), meaning that we must deduce the characteristics of the elementary electromagnetic particles that constitute the fundamental level of objective reality from what we can indirectly detect and understand from the behavior of atoms, and from the behaviour of the elementary particles that can be separated from them; *i.e.* electrons whose stabilization far from the nuclei determines the volume of space occupied by atoms, and from the behavior of protons and neutrons that constitute their nuclei by occupying smaller volumes; as well as from the behavior of the electromagnetic energy which is emitted or absorbed by these elementary particles during their transitions between the various stationary action equilibrium states in which atoms stabilize at the atomic level.

Finally, the means we have at our disposal to observe the behavior of atoms and their separable elements is precisely the electromagnetic energy which is emitted or absorbed during these stationary action equilibrium states variations, and whose *infinitesimal granules, i.e.* these localized electromagnetic photons coming from all objects in our surrounding, either directly from these objects or detected through our powerful microscopes and other sensing devices, that excite electrons from the atoms forming the photosensitive cells in our eyes, an excitation which is then progressively transmitted along our optic nerves to the brain that continuously updates the images of which we become aware from our environment and that we analyze to understand it [25].

These localized electromagnetic photons that can excite electrons sufficiently in the cells of our eyes for their arrival to be progressively signaled all along the optic nerve, can be of very variable intensities, and above a certain intensity level, succeed in separating the electrons from the atoms in our environment, and this is what allows us to study their separate behavior as well as that of the constituents of atomic nuclei, namely protons and neutrons, which can be completely separated from their electronic escorts and studied separately in the case of simple atoms such as hydrogen or helium atoms.

What was preventing us up to now from becoming as comfortable treating electromagnetic energy as being granular, that is quantized, at the subatomic level, as we are handling it as continuous electromagnetic waves at our macroscopic level, is that since about a hundred years, the quantized aspects of the subatomic level have been considered being the exclusive domain of Quantum Mechanics (QM), but that QM still has not been fully harmonized with Maxwell's electromagnetic equations, that successfully handle electromagnetic energy as a continuous wave at our macroscopic level; in other words, that treats it as a fluid, which is an incomplete harmonization that was clearly highlighted by Feynman, who was the last researcher to attempt this reconciliation in the mid 20th century, as evidenced by this quote from his *"Lectures on Physics"* [26]:

> *"There are difficulties associated with the ideas of Maxwell's theory which are not solved by and not directly associated with quantum mechanics...when electromagnetism is joined to quantum mechanics, the difficulties remain".*

As put in perspective in a recent article ([11], see also Sections 3.9 to 3.12), all current theories mathematically treat macroscopic masses as if they had no internal granular structure, that is, as if they were made of a continuous substance uniformly spread within their whole volume, and even Quantum Mechanics currently treats the electron energy as if it was uniformly spread in the same manner within the volume defined by the Schrödinger equation. The reason for this is that the internal electromagnetic structure of the energy making up the mass of each elementary particles of which all macroscopic masses are made, such as the electron, as well as the internal electromagnetic structure of those making up the inner structures of the protons and neutrons, that constitute the nuclei of all atoms in the universe, have not yet been clearly established; and that the momentum energy as well as the energy causing the increase of the transverse magnetic field of accelerating particles have not yet been mathematically separated from the energy of which their rest masses are made.

Recently, however, new developments have made it possible to establish a coherent internal subatomic electromagnetic structure for localized electromagnetic photons and for all elementary electromagnetic particles in accordance with Maxwell's equations, which finally makes it possible to find natural the perception that all atoms are made at the subatomic level of separate and localized elementary particles stabilized in various states of stationary action electromagnetic resonance states and that free moving electromagnetic energy is quantized at the subatomic level, even if we treat it as a continuous wave at our macroscopic level.

1.3. Two recent major breakthroughs

Already in the 1930's, Louis de Broglie proposed the hypothesis of a possible potentially quantized internal structure for localized electromagnetic photons at the subatomic level that would remain conform to Maxwell's equations, but whose elaboration, by his own admission, seemed not to be possible in the restricted frame of the 4-dimensional geometry of Minkowski's space-time:

> *"... la non-individualité des particules, le principe d'exclusion et l'énergie d'échange sont trois mystères intimement reliés : ils se rattachent tous trois à l'impossibilité de représenter exactement les entités physiques élémentaires dans le cadre de l'espace continu à trois dimensions (ou plus généralement de l'espace-temps continu à quatre dimensions). Peut-être un jour, en nous évadant hors de ce cadre, parviendrons-nous à mieux pénétrer le sens, encore bien obscur aujourd'hui, de ces grands principes directeurs de la nouvelle physique." ([27], p. 273).*

Translation:

> *"... the non-individuality of particles, the exclusion principle and exchange energy are three intimately related enigmas; all three are*

tied to the impossibility of exactly representing elementary physical entities within the frame of continuous three dimensional space (or more generally of continuous four dimensional space-time). Some day maybe, by escaping from this frame, will we better grasp the meaning, still quite cryptic today, of these major guiding principles of the new physics."

Two recent developments, however, made it possible to elaborate this internal electromagnetic structure of the localized photon proposed by de Broglie in full conformity with Maxwell's equations, and to eventually observe that all stable massive and electrically charged elementary particles of which all atoms are made at the subatomic level can also be described in the same Maxwell compliant manner.

The new light shed by these recent developments on the nature of fundamental electromagnetic energy then made it possible to refocus according to this new perspective the bulk of the conclusions drawn in the past from all experimental data collected to date about the subatomic level. These refocused conclusions were then explained in about twenty separate articles, each of which analyses a specific aspect of the issue, most of which will be given in reference during this final synthesis.

1.4. The first major breakthrough

The first of these two breakthroughs was the elaboration of a more extensive geometry of space, based on the three-way orthogonal relationship that Maxwell related to the three fundamental aspects of electromagnetic energy of which light is made at the subatomic level, namely its electrical and magnetic aspects perceived as being perpendicular to each other and mutually inducing each other into a standing cyclic transverse oscillation mode of the energy that these fields measure with respect to the direction of motion in vacuum of this transversely oscillating electromagnetic energy in space, that is, a direction of motion of this energy which is perpendicular to the direction of the stationary transverse oscillation of the energy represented by the two fields (See **Figure 1.1**).

The trispatial geometry (see **Figure 1.3**) required to develop the LC equation derived from the de Broglie hypothesis ([15], [8] Chapter 6) in accordance with

Maxwell's interpretation (**Figure 1.1**) was formally presented at the event Congress-2000 in July 2000 at St Petersburg State University [28].

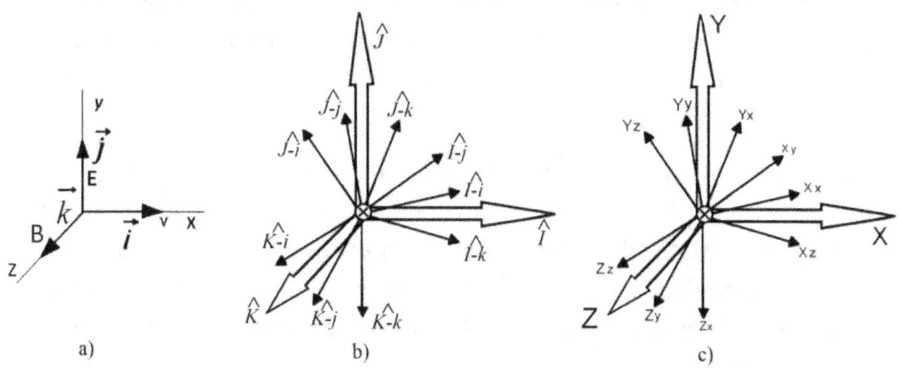

Figure 1.3: Major and minor vectors sets applicable to the trispatial geometry.

This expanded space geometry at the subatomic level is fully described in Reference ([10], See also Chapter 2), but can be briefly summarized as follows. The method consists in geometrically expanding each of the 3 standard linear electromagnetic vectors *i*, *j* and *k* (**Figure 1.3-a**), applicable to normal space, transforming them into 3 fully developed 3D vector spaces of their own (**Figure 1.3-b**), each of these three spaces, now identified as spaces X, Y and Z (**Figure 1.3-c**), each space remaining perpendicular to the other two and all three remaining connected via their common punctual origin.

This common centre can now be understood as serving as a passage point located at the centre of each localized electromagnetic quantum at the subatomic level, through which the *energy substance* of the particle would be free to circulate between the three spaces as if between communicating vessels, so as to allow the establishment of a stationary transverse oscillation of half the particle energy between its *E* and *B* aspects between the two YZ-spaces, as well as an equal sharing of the total energy of the particle between the transversely oscillating energy half-quantum of the *E* and *B* fields within the YZ-transverse-dual-space-complex, and the unidirectional energy half-quantum of the momentum of the particle residing in X-space.

To mentally visualize the motion of energy in this trispatial geometric complex of 9 mutually orthogonal dimensions, it suffices to imagine each of the 3 sets of minor vectors *i*, *j* and *k* of **Figure 1.3-b** as if they were the folded ribs of 3 metaphorical umbrellas. This allows any of them to be mentally opened at will one at a time up to full orthogonal expansion to observe and mathematically describe the behavior of energy in this fully deployed 3D space during each phase of its oscillating motion. **Figure 1.3-b** and **1.3-c** show the dimensions of the 3 spaces only half-deployed to allow a clear and unique identification of each of the 9 resulting internal orthogonal axes.

1.5. The second major breakthrough

The second major development occurred a few years later, in 2003, when Paul Marmet published an important article describing a newly perceived relation between the progressive increase of the intensity of the transverse magnetic field of an accelerating electron and the simultaneous increase of its transversely measurable mass [29], that then allowed clearly distinguishing between the variable energy of the electron momentum that also increases during its acceleration, and the also variable energy of its transverse magnetic field, and also to clearly separate these two variable energy quantities from the invariant amount of energy constituting the electron rest mass as described in an article published in 2007 in the same "*International IFNA-ANS Journal*" at Kazan State University ([30], [8] Chapter 4).

This discovery then allowed observing that all charged elementary particles constituting atoms have the exact same internal electromagnetic LC structure in this expanded space geometry, each one being accompanied by an amount of carrying energy made of momentum energy and transverse magnetic field energy, which is structured in a manner identical to the internal electromagnetic structure described by the LC equation previously developed to account for localized double-particle photons as hypothesized by de Broglie ([15], [8] Chapter 6) ([31], [8] Chapter 11) ([32], [8] Chapter 14) ([33], [8] Chapter 12), which then allowed establishing their respective trispatial LC equations, as summarized in Reference ([10], See also Chapter 2), as we will see further on.

Let us note here that this internal electromagnetic LC structure is also applicable to all of the electrically charged elementary electromagnetic particles constituting the complex unstable particles, be they electrically neutral or not, such as pions, kaons and other ephemeral complex particles resulting from destructive scattering between elementary particles ([34], [8] Chapter 19).

We will study here however only the stable particles making up the scatterable structure of the set of atoms that can be found in the periodic table and of their nuclei, as well as positrons and free moving electromagnetic photons, because all of the unstable partons generated via destructive scattering play no role whatsoever in the establishment and stability of the universe, since they all almost instantly decay by releasing their excess energy in well known sequences of stages [35], until all that remains of them is one or other, or many of the very restricted set of stable electrically charged and massive elementary particles making up all atoms ([34], [8] Chapter 19).

But attention must first be given to a typographical error in Equation (M-7) of Marmet's article that renders the seamlessness of his derivation difficult to perceive. For his unbroken sequence of reasoning to be made clear, his derivation down to Equation (M-7) from the Biot-Savart equation will be fully detailed here. The remainder of his derivation down to Equation (M-23) remains easy to follow directly in his article [29] and is also clearly explained and analyzed in another recently published article ([10], See also Chapter 2).

Although the second part of his article starting with Section 7 elaborates a personal hypothesis on a possible inner structure of the electron, which is of course subject to discussion, the first part of his article is in no way hypothetical, but rather elaborates a mathematically seamless derivation from the Biot-Savart equation, itself established directly from experimental data that can easily be re-obtained at will, that leads to the establishment of a new Equation (his equation M-23) that effectively seems to leave no doubt, quoting Marmet himself , that: *"the increase of the so-called relativistic mass* [of an accelerating electron] *is in fact nothing more than the mass of the magnetic field generated due to the electron velocity"* [29]:

$$\frac{\mu_0 \left(e^-\right)^2}{8\pi} \frac{1}{r_e} \frac{v^2}{c^2} = \frac{M_e}{2} \frac{v^2}{c^2} \tag{M-23}$$

To avoid any confusion in the numbering of equations in the present article, all equations quoted from Marmet's paper will be prefixed with "M-" followed by the actual number of this equation in the original paper [29], so readers can directly relocate them in his paper.

Equation (M-23) suggests numerous possibilities that never were considered before, the most important of which is that it highlights an inconsistency between the Special Relativity Theory (SR) and electromagnetism that could not be noticed otherwise, because the very idea that the energy that progressively increases the transverse magnetic field of an accelerating electron, as calculated with the equations of electromagnetism, could be the same energy measurable as its transverse mass progressively increasing with velocity, as calculable with the equations of relativistic mechanics, is absent from SR for a reason that will be highlighted later.

The first clue suggesting the possibility that a single quantum of energy might be responsible at the same time for the increase of the electron's transverse magnetic field and for the increase of its transversely measurable relativistic mass, is established by the well-known fact that the magnetic field, as measured around a wire conducting a stable electric current, which is of course made of electrons circulating all at the same velocity and in the same direction in the wire, is oriented perpendicularly, that is, transversely, with respect to the direction of motion of the electrons, which is what the Biot-Savart law accounts for, as put into perspective by Marmet at the beginning of his article [29].

An important point must already be highlighted regarding the habit acquired since Maxwell to think of the familiar three-way orthogonal relationship of electromagnetic energy as involving electric and magnetic *fields* perpendicular to each other, that would be at the same time perpendicular to the direction of motion of the energy.

It is a fact seldom mentioned in reference works that the concept of the *electric field* introduced by Gauss was only meant to *conceptually represent* the Coulomb interaction in an *idealized geometric and mathematical manner* as diminishing omnidirectionally towards zero at infinity, according to the inverse square of the distance rule, from a maximum value located at the point in space where the single test charge remaining in the Coulomb equation would be located when the second charge is removed from the equation, as highlighted in a recent article [25]. This idealized concept was then also conceptualized geometrically and mathematically to represent in the form of a *magnetic field* the magnetic aspect of electromagnetic energy.

It will therefore be important for the remainder of this analysis to keep in mind Gauss's original intention that these *fields* should be considered only as *idealized geometrical and mathematical tools* only meant to *represent* the real energy which is deemed to physically exist, and that it is the electromagnetic energy itself that physically exists that would physically self-structure, so to speak, according to this dual perpendicular pattern resulting from its transverse electromagnetic oscillation, that is, an oscillation which is transversely oriented with respect to the unidirectional momentum energy that sustains its motion in space.

It follows that the transverse energy itself that Marmet's derivation identifies as simultaneously accounting for the transverse magnetic field increase and the measurable transverse relativistic mass increase [36] of the accelerating electron, can therefore only be oriented perpendicularly to the direction of motion of the electrons whose circulation generates the stable current measurable via the Biot-Savart equation.

This of course means that the energy that supports the increasing momentum of an accelerating electron, that can be calculated with relativistic mechanics equation $\Delta K = \gamma m_0 v^2/2$, can in no way be the same as the energy that perpendicularly supports its increasing transverse magnetic field that can be calculated by means of the Biot-Savart equation, the latter now presumably corresponding to the energy of the transverse relativistic mass increment computable with the relativistic mechanics equation $\Delta E = \Delta mc^2 = (\gamma m_0 c^2 - m_0 c^2)$, because it is physically and vectorially impossible for a single energy quantum to move in both of these two perpendicular directions simultaneously, and also because the total amount of only one of these two energy quantities is insufficient to singlehandedly account for the simultaneous energy increase of both its longitudinal momentum and of its transverse magnetic field at any given velocity.

On the other hand, Maxwell's first equation (Appendix B), which is in fact Gauss's equation previously mentioned for the electric field, and that reconverts to the simple Coulomb equation when a second charge is introduced in the *idealized field* of the test charge, reveals that the total amount of energy induced in each accelerating charge amounts to twice the energy of the longitudinal momentum $\Delta K = \gamma m_0 v^2/2$, or alternatively, to twice the energy of the transverse relativistic-mass/magnetic-field increment $\Delta E = \Delta m_m c^2$. More to the point, this reveals that both amounts are always equal by structure and that this sum can only be made of their simultaneously induction, in which ΔE_{total} also accounts for the accelerating electron transverse magnetic field increment, both quantities thus making up the total amount of energy required to account for the simultaneous increase of the velocity and of the related transverse magnetic field, that is, $\Delta E_{total} = \Delta K + \Delta m_m c^2 = \gamma m_0 v^2/2 + (\gamma m_0 c^2 - m_0 c^2)$, as demonstrated in Reference ([10], See also Chapter 2).

We should therefore rather speak in reality of two energy *half-quanta* constituting a single quantum of induced energy. The fact that this total quantum of energy calculated with the Coulomb equation varies in an infinitesimally progressive manner uniquely as a function of inverse of the distance separating two charged particles, also demonstrates that this energy varies adiabatically, and this, uniquely as a function of the inverse of the distances separating all charged particles from each other on account of to the Coulomb interaction, whether they are moving or not.

An additional clue supporting the conclusion that these two energy half-quanta have to exist simultaneously, is that to even be able to calculate the ΔB magnetic field increment related to any velocity of an accelerating electron with the generalized form of Marmet's Equation (M-7) established in Reference ([30], [8] Chapter 4), it is the wavelength of this double amount of energy given by the Coulomb equation that must be used to obtain this correct ΔB value of the transverse magnetic field increment of the moving electron, which will be demonstrated with Equation (1.9) further on.

1.6. Historical context of the development of the theory of Special Relativity (SR)

But the very fact that these two energy half-quanta are always equal in quantity initially induced confusion in the community in the absence of this new information available only since Marmet's recent derivation. This confusion led to considering that a total amount corresponding to only one of these two half-quanta was induced during the electron relativistic acceleration process, which gave rise to a famous disagreement among the theoreticians of the beginning of the 20th century.

For example, Minkowski [2], Lorentz [37] and Einstein [38] related this half-quantum of energy strictly to momentum, a conclusion which is an integral part of the Theory of Special Relativity (SR), while Abraham [39], Poincaré [40] and Planck [41] related the half-quantum of measured motion energy strictly to an increase in the transversely measurable mass.

1.7. The conclusion of Minkowski, Lorentz and Einstein

By consulting a famous article by Max Planck dating from 1906 [41], it can be noted that he refers to the energy constituting the mass of a moving electron $E=\gamma m_o c^2$ by the terms "*lebendige Kraft*" (See his comment following Equation 8, page 140 of his text, identifying this energy by the term "L"), which is translated in the fundamental physics community by the terms "*kinetic force*" (or "*vibrating force*" or "*live force*" for a literal translation from German), which puts in perspective that at the beginning of the 20th century, the relation between the concept of *force*, such as the force calculable with the Coulomb equation or with the fundamental mass acceleration equation $F=ma$, that we conceptualize as having dimensions *joules per meter* ([14], [8] Chapter 13), and the concept of *energy induced by the Coulomb force*, which is obtained by multiplying the Coulomb force by the distance between two electric charges, and that we conceptualize as being in *joules* only ([14], [8] Chapter 13), was not yet clearly established; these two concepts apparently not yet having been clearly differentiated. The only reference to momentum in his text is "*Impulskoordinaten*" ("*momentum coordinates*"), which he does not associate with the energy that supports it in context of the ongoing debate at that time, and this at the very historical moment when this debate about the introduction of the SR Theory was raging.

By comparison, in the German fundamental physics community today, the momentum "*Impuls*" is immediately conceptualized as a quantity of kinetic energy "*kinetische Energie*" moving in a specific vectorial direction, as in the physical communities of other languages. Few today are fully aware that at the beginning of the 20th century, the greatest advances in fundamental physics were made in Europe, and that the original articles were written mainly in German, but also in French and Italian, and that some of these founding articles have still not been formally translated to English, contrary to popular belief, and some very belatedly. For example, the text of a seminal presentation by Herman Minkowski from 1907, "*Das Relativitätsprinzip*", was only very recently translated to English in 2012 by Fritz Lewertoff [2]. Practically all of Louis de Broglie's writings, whose complete work has just been translated to Russian, have not yet been translated to English. It is therefore important to consult formal articles in their original language, to ensure that the translated versions are accurate, and more importantly to correctly put in perspective the lesser extent of the established knowledge pool at the time of their writing and on which they were grounded.

Analyzing Lorentz's article of 1904 [37], that introduced the concept of relativity by incorporating the γ factor into the equations of classical mechanics, which is what prompted Planck to write his 1906 paper [41] previously quoted, it can be seen that the concept of the Coulomb force is clearly defined, but that the energy of the relativistic momentum of the electron is calculated in the manner that intuitively comes to all our minds initially; that is, by simply adding the γ factor to Newton's initial non relativistic classical kinetic energy equation $K=m_o v^2/2$, but that he does not modify this equation to incorporate the half-quantum of transverse energy that supports the corresponding increment of its magnetic field, as described in Reference ([42], [8] Chapter 5), or alternatively, that he does not multiply the force obtained by means of the Coulomb equation by the distance separating the two charges to obtain the total amount of energy adiabatically induced in each charges by the Coulomb interaction at this distance, as described in Reference ([10], See also Chapter 2).

We should therefore become fully aware that if two of the greatest discoverers of the time, Planck and Lorentz, had not become aware of the ontological relation now obvious to us between the Coulomb interaction and the induction of kinetic energy in charged particles, and of the relation between this electromagnetically induced energy and the kinetic energy causing massive bodies to move, from the classical/relativist mechanics perspective, macroscopic bodies whose masses can only be exclusively made of the sum of the masses of these electrically charged elementary particles, it necessarily means by extension that this relationship was not yet clearly established in the whole scientific community of the time, as unexpected as this may seem today.

It remains however astonishing that the great discoverers of that time were able to establish so precisely the equations of classical/relativistic mechanics without having benefited from the hindsight now provided us by a further century of experimentation, which now makes it possible to clearly perceive this relation between the so-called *Coulomb force*, obtained by multiplying the unit charge of the electric field equation established by Gauss $E= e/4\pi\varepsilon_0 d^2$ [17] by a second charge e, which acts according to the rule of the inverse square of the distance between electric charges $1/d^2$, that is $F=e\cdot E= e^2/4\pi\varepsilon_0 d^2$ ([25] Equation (4)), and the amounts of *adiabatic kinetic energy* ([43], [8] Chapter 2) that this force induces in these electric charges as a function of the simple inverse of the distance separating them *1/d*, that is $E=d \cdot F = e^2/4\pi\varepsilon_0 d$ ([25] Equation (4)), which are concepts that seemed difficult to clearly correlate through the fog of uncertainty that still pervaded the relations between these electromagnetic concepts that were then not in process of being methodically explored, and that still are not today (see following Section), and the classical concept of *mass*, that belonged to the domain of classical mechanics, and that still was considered as unrelated to electromagnetism at that time, except by Einstein himself, but without relating it to gravitation at that moment, as we will soon see in coming Subsection 1.7.1 and 1.7.3.

This is what explains why the concept of *force* was not specifically incorporated to SR to justify the increase in energy of a moving or accelerating mass, and also why the very notion of *force* is simply absent from the theory of General Relativity (GR), in which it is replaced as the ontological cause of the existence of energy by an inertial motion of massive bodies caused by an assumed *curvature* of *space-time*, which prevented the Coulomb equation, which is based on the concept of a *force* associated with the acceleration of electrically charged particles, from being conceptually associated with the acceleration of the electron *mass* from this perspective, because no direct connection is made in this theory between the concept of *classical mass* and the fact that all macroscopic massive bodies can only be made of electrically charged massive elementary particles ([11], See also Chapter 3), as will be put into perspective later.

As strange as this may seem, more than one century after Kaufman's defining experiments with electrons accelerating to relativistic velocities [36], no concept of an increase of the magnetic field of the accelerating electron mass exists in SRT, which makes it seem normal according to this theory that only the momentum energy half-quantum would be increasing with velocity, that is, a velocity apparently due to a theoretical *inertial acceleration*.

1.7.1 The interesting case of Albert Einstein's claim regarding electromagnetism

It may seem paradoxical, as just asserted, that the concept of force seemed not to have been incorporated to SR for the reason that the classical concept of mass was considered unrelated to electromagnetism at that time, since we just learned from Reference [5], that Einstein apparently indirectly but correctly understood the relation between the force of acceleration that applies to the invariant charge of the electron, and the force of acceleration that applies to its invariant rest mass, as demonstrated by his Equation (2), previously mentioned in the introductory Foreword:

$$(2) \qquad m\frac{\mathrm{d}^2 x}{\mathrm{d}t^2} = e\mathbf{E}_x \qquad\qquad ([5], \text{p. } 143)$$

Indeed, $F = m\dfrac{\mathrm{d}^2 x}{\mathrm{d}t^2}$ happens to be one of the numerous mathematical representations of the fundamental acceleration equation $F=ma$, as emphasized in Reference ([25] Section 27), and $F = e\mathbf{E}_x$ is one of the numerous mathematical representations of the Coulomb equation as put in perspective in Section 1.7. See also Reference ([25] Equation (4), reproduced here for convenience:

$$\mathbf{F} = e\mathbf{E} = \frac{e^2}{4\pi\varepsilon_0 r^2} \qquad\qquad ([25] \text{ Equation (4)})$$

due to the fact that the symbol for the electric field (E) was defined by Gauss as being equal to the following definition by removing one of the charges from the Coulomb equation:

$$\mathbf{E} = \frac{e}{4\pi\varepsilon_0 r^2} \qquad\qquad ([25] \text{ Equation (3)})$$

Obviously, when the missing charge is re-introduced as Einstein did in Reference ([5] Equation (2)), the complete Coulomb force equation is re-established.

Now, Einstein does not clarify how he initially derived this equality between these two proven force equations, so they end up *de facto* axiomatically being established as a *single force* applicable simultaneously to the rest mass of an electron and to its invariant charge. So it would seem that he really posed this equality as an axiom in 1910, which was a habit of his in establishing grounding foundations for his reasonings, such as the grounding axioms of his previously established Special Relativity theory. See Sections 3.4 and 3.5 about this issue.

But it so happens that a mathematical derivation demonstrating that all classical force equations are only alternate representations of Newton's $F=ma$ acceleration equation in Reference ([44], [8] Chapter 7), clearly demonstrates that Einstein's assumption in posing Equation (2) was completely justified.

Now, from Einstein's introducing comment to his Equation (2), as quoted in the Foreword, although he related the electron charge to the E-field, it is far from obvious that he relates any more clearly than his colleagues the E-field symbol with the detailed sub-definition that Gauss meant it to represent, that is, the Coulomb equation minus one charge.

By comparison, see with Equation (3.3) how the same Newton's acceleration equation is directly set as equal to the Coulomb equation in standard physics introductory textbooks, this time without relating it to the Gauss E-field equation as combined in the Lorentz equation with the charge that is missing to re-obtain the Coulomb equation, which is a state of fact that prevents most students from seeing the direct relation between the Gauss equation and the standard Coulomb equation.

Actually, from decades of discussions with hundreds of physicists, I observed that very few of them usually conceptualize the E-field as being so directly related to the Coulomb equation, even when a second charge is involved, such as in the Lorentz force Equation ($F = q(E + v \, x \, B)$), and it would seem that this was already the case at the beginning of the 20th century, since physicists typically seem to generally prefer conceptualizing electromagnetism from the potential fields perspective.

This is moreover confirmed by the very text of Einstein's article [5] when he states:

> *"...on s'habitua à considérer les champs électrique et magnétique comme des entités dont l'interprétation mécanique était superflue. On en vint ainsi à regarder ces champs dans le vide comme des états particuliers de l'éther, n'exigeant pas une analyse plus approfondie."*

Translation:

> *"...the electric and magnetic fields eventually came to be considered to be entities whose mechanical interpretation was superfluous. This led to the view of these fields in vacuum as particular states of the aether that did not require further analysis"*

Nowhere in his article is the Coulomb Law mentioned or associated with the electric field or electrically charged particles. The motion of electric charges is mentioned, in agreement with H.A. Lorentz, as being strictly due to their interaction with the electric field:

"Une particule chargée en mouvement par rapport à l'éther est assimilable à un élément de courant; les actions du champ électromagnétique sur la particule et les réactions de cette dernière sur le champ sont les seuls liens qui lient la matière à l'éther. Dans celui-ci, là où l'espace n'est pas déjà occupé par une particule, les intensités du champ électrique et magnétique sont exprimées par les équations de Maxwell pour l'éther libre, si l'on suppose que les équations sont rapportés à un système d'axes immobile par rapport à l'éther."

Translation:

"A charged particle in motion relative to the aether is like an element of current; the actions of the electromagnetic field on the particle and the reactions of the latter on the field are the only bonds that bind matter to the aether. In the latter, where space is not already occupied by a particle, the intensities of the electric and magnetic field are expressed by Maxwell's equations for the free aether, assuming that the equations are related to a system of axes which is immobile relative to the aether."

All considered, it can be concluded that the direct identity that Einstein perceived between Newton's fundamental acceleration equation applied to the rest mass of an electron and the acceleration equation of the unit charge of the electron submitted to an electric field, probably is what eventually convinced him that gravitation has to follow the pattern of electromagnetism.

1.7.2. The surprisingly incoherent objection of Archibald Wheeler

Let us now have a look at the justification quoted in the Foreword that Wheeler provided for his refusal to consider Einstein's conclusion as a possibly valid avenue of research, apparently also self-appointing himself as speaking for the whole community, without anyone raising a protest: *"This thesis, we cannot accept, and the community of physics, quite rightly, does not accept."* ([7], p. 391).

Unfortunately, he did not provide a reference to the specific text from Einstein in which the latter would have *"claimed"* that gravitation follows the pattern of electromagnetism. So this still is a search in progress at the moment of publication of this book.

Quite unexpectedly, just before formulating his rejection of the possibility that gravitation might follow the pattern of electromagnetism, Wheeler opposes a completely invalid version of the Coulomb equation:

$$\mathbf{F}_{electr} = e_1 e_2 / r^2 \qquad\qquad \text{([7] Equation (7.1.2))}$$

to the valid gravitational equation:

$$\mathbf{F}_{grav} = -G m_1 m_2 / r^2 \qquad\qquad \text{([7] Equation (7.1.2))}$$

and then concluded that this visibly erroneous comparison completely disqualifies electromagnetism as a potentially promising avenue of investigation in search of a possible pattern that could relate it to gravitation.

This version of the Coulomb equation is invalid for the simple reason that Wheeler forgot, or, is it even conceivable, <u>did not know</u>, that to be dimensionally valid, this equation must involve the Coulomb proportionality constant (see Equation (2.18) and the related comment, in Chapter 2):

$$k_e = \frac{1}{4\pi\varepsilon_0} \text{ (Electrostatic Coulomb Constant)} \qquad\qquad (2.18)$$

Even at face value, the dimensions of the force obtainable from the erroneous Wheeler version of the Coulomb equation are obviously inconsistent, resolving to *coulombs squared per meters squared* (C^2/m^2), whereas it is well established that a force can only be expressed in newtons (N), that resolve to their elementary dimensions *joules per meter* (j/m), which are dimensions that can be obtained from the Coulomb equation only if the Coulomb constant is involved: $F_{electr} = k_e e_1 e_2 / r^2$, and which are identical to the dimensions of the force obtained from the correctly posed gravitational equation.

It is consequently quite remarkable and unexpected that such a blatant error in a so popular reference work [7], that was used as a justification to refuse to explore such a so fundamentally important field as electromagnetism for a possible relation with gravitation, seems not to have drawn attention in the physics community, particularly in light of, and in opposition to, the specific recommendation from the most famous physicist of the 20th century, and also that this type of error in such a simple equation is likely to draw the immediate attention of anyone with minimal mathematical proficiency.

1.7.3. The solution that Einstein possibly was searching for

A point of high interest regarding Einstein's search to associate gravitation to electromagnetism comes up with regard to the correctly formulated gravitational equation referred to by Wheeler, that is, ([7] Equation (7.1.2)) previously mentioned.

Having directly related both the Lorentz electrostatic force equation with Newton's fundamental acceleration equation by means of Equation (2) ([5], p. 143), no doubt was Einstein looking to also directly relate the gravitational equation to these first two force equations.

It so happens that this direct equality between these three equations was mathematically established in Reference ([44], [8] Chapter 7, Equation (7.48)) with respect to the invariant rest mass and invariant charge of the electron, in line with Einstein's equality established between the first two classical force equations in his Equation (2). Moreover, it was demonstrated in the same reference that all 5 classical force equations can be derived from each other by means of the generalized form of the Coulomb equation that was developed in Reference ([30], [8] Chapter 4, Equation (4.11)):

$$F = G_p \frac{M_p \bullet m_e}{r_0^{\,2}} = k \frac{e^2}{r_0^{\,2}} = e\mathbf{vB} = e\alpha\mathbf{E} = m_e a = 8.238721807\mathrm{E}-08\,\mathrm{N} \qquad ([44]\ \text{Equation (48)})$$

This common denominator that the Coulomb equation happens to be in allowing to link all classical force equations *is what allows to mathematically link the adiabatic energy permanently induced in all the elementary charged particles by the Coulomb force to all classical force equations, and consequently to gravitation*, in Sections 1.26 and 1.27 below, as established in Reference ([43], [8] Chapter 2).

1.8. The Conclusion of Planck, Poincaré and Abraham

As mentioned previously, Abraham [39], Poincaré [40] and Planck [41] related the half-quantum of measured motion energy strictly to an increase in the transversely measurable mass, without relating it however in any way to the simultaneous increase of the related transverse magnetic field. From this perspective, the momentum of a moving mass does not have a physical existence, but is considered as an impulse propagating in an underlying aether that would propel the mass, which also makes it seem normal from this second perspective that only the energy half-quantum of the transversely measurable mass increases with velocity.

This disagreement between the position of Einstein, Minkowski and Lorentz on the one hand, and that of Poincaré, Abraham and Planck on the other, is still the object of endless discussions in the community. In both cases, no relation is established with the double amount of energy revealed by the Coulomb equation as being ontologically induced simultaneously by the Coulomb interaction in the accelerating electron; and neither of these solutions allows even suspecting that these two half-quanta could be increasing simultaneously.

Consequently, gaining a clear awareness of the mandatory simultaneousness of the existence of both of these two energy half-quanta perpendicularly oriented with respect to each other, in light of Marmet's discovery and in relation with the Coulomb equation, is therefore required for a complete harmonization of classical/relativistic mechanics and electromagnetism to be realized.

1. 9. The Absolute Axiomatic Principles

Let us return for a moment to the previously mentioned "*fog of uncertainty*" that surrounded the concepts of the Coulomb force and the energy induced by this force as the theory of Special Relativity was being developed at the beginning of the 20th century.

Throughout history, before the extent of the momentary accumulation of knowledge about Nature made it possible to identify absolute constants in Nature on which theories could be grounded to explain the identifiable processes observable in objective reality, the method used to ground these theories consisted in establishing absolute axiomatic *principles* to be used as stable references to firmly ground rational explanations about the nature of energy, mass, electric charges, etc. These principles eventually became *idealized dogmas* that the scientific community adopted as being reliable references to ground the theories that were in process of being developed, such as the Principle of conservation of energy, the Pauli Exclusion Principle, the Principles of stationary action and of least action, etc.

Some of these Principles are *positive* idealized Principles, such as the Principle of conservation of energy, that bar all possible exceptions, but that do not actively discourage research as to possible limitations of their reach or of even the very validity of a principle with respect to its applicability to physical reality, that may have been less well understood when it was initially formulated.

Indeed, in the case of this last principle, for example, the current extent of knowledge allows now to better define its reach with respect to physical reality, because we can observe that the Principle of conservation of energy remains valid for a system as long as this system previously stabilized in some stationary action equilibrium state returns to this state after having been disturbed, but that if it is led to vary in such a way as to stabilize axially into a less energetic or in a more energetic stationary action state than the initial state, this change can only be adiabatic in nature ([43], [8] Chapter 2).

This is precisely the case for the space probes that were taken away from Earth and launched on least action escape trajectories from the solar system, for example, as we will see later ([45], [8] Chapter 16) [46] [47] [48]. When such systems stabilize in such a new state of stationary action axial equilibrium, the Principle of energy conservation applies again, but with reference to this new state of stationary action axial equilibrium. Indeed, the masses of which these probes are made will never return to the state of stationary action axial equilibrium that they had before launch.

In reality, all stationary action states allowed in objective reality are part of a hierarchy of axially distributed stationary least action electromagnetic equilibrium states, ranging from the stationary states of the subatomic order of magnitude to those of the astronomic order of magnitude, whose detailed hierarchical correlation remains to be completely established, and the only way for an elementary particle or larger mass to move axially from one of these stationary equilibrium state to another is by means of a least action trajectory involving an adiabatic change in its carrier energy. This hierarchy of stationary states will be discussed further on, but let us return for now to the main theme of this section, which is the set of historically established absolute axiomatic principles.

Among the set of historically established *positive* axiomatic dogmas, however, is one, the *de facto* rejected concept of *action-at-a-distance*, also derogatively referred to as *spooky-action-at-a-distance*, which is universally and unjustifiably associated with the Coulomb's so-called *force*, which is a *negative* and *absolute* dogma, in the sense that it actively discouraged any research in the community in trying to study and understand the nature of the Coulomb interaction, despite the fact that it directly underlies Maxwell's first equation, that is, Gauss's equation for the electric field as previously described, which is universally accepted as valid.

The misunderstanding that apparently led to the very idea of a so-called *action-at-a-distance* in reference to the Coulomb *force* seems to have been that this so-called *force* was associated to the concept of an *attraction*, as defined in Newton's macroscopic gravitational theory, instead of having been associated to a *process of energy induction, half of which provides unidirectional momentum* in electrically charged particles at the subatomic level, and that an assumed *attraction* between charged particles was wrongly considered as being due to an *attractive force*, instead of being understood as a motion *propelled by some unidirectional momentum energy* of an electrically charged particle towards another electrically charged particle of opposite sign; and that an assumed *repulsion* wrongly interpreted as being due to a *repulsive force* between electrically charged particles of same sign, turns out to be in reality a motion of an electrically charged particle away from another electrically charged particle of same sign *propelled by some unidirectional momentum energy*, with *no force* whatsoever being involved in the process, as analyzed in Reference ([11], See also Section 3.17).

The concept of Coulomb interaction having now been summarily formulated in a manner more in line with reality, and in order to distance ourselves from the concept of Newtonian *force*, which is useful at the macroscopic level, but is deceptive when dealing with massive and charged elementary particles at the subatomic level, the expressions *Coulomb interaction* will generally be used for the remainder of this article instead of the misleading expression *Coulomb force*.

A hundred years now after Lorentz, Planck, Einstein, de Broglie and Schrödinger, to name only a few of the extraordinarily dedicated scientists of the time who revolutionized fundamental physics at the beginning of the 20th century, it seems that we know enough now about the subatomic level to do away with such absolute axiomatic principles and dogmas, by either clearly identifying the physical limits of their applicability, as in the case of the Principle of conservation of energy, or by simply doing away with those that ultimately turn out to just having been misguided impediments to research, like the concept of *action-at-a-distance*, due to insufficient initial information having been available as to the actual possible nature of the Coulomb interaction, for example, that we now know is the actual cause of the simultaneous adiabatic induction of both perpendicular energy half-quanta in all charged elementary particles in existence, that is, a Coulomb interaction whose nature still remains to be clearly understood.

1.10. Inappropriate names given to some processes and states

The very names given in the past to some stable observed characteristics and processes of elementary particles, before the electromagnetic nature of the energy of which their invariant rest masses is made was understood, also heavily contributed to the persistent confusion in the community as to the real nature of these characteristics and processes.

For example, the lower limit of integration of the energy of the rest mass of the electron obtained by means of the spherical integration mathematical means, was quite inappropriately named *the electron classical radius*, symbolized by r_e, which constantly tends to cause many researchers *to think* of this value as possibly representing the true physical radius of the electron mass, in the classical mechanics sense ([30], [8] Chapter 4).

Another much more insidious misnomer is the term "*spin*" chosen to refer to the relative magnetic polarity of mutually interacting electrons and of their interaction with the electromagnetic subcomponents of nucleons, that induces the quite inaccurate beliefs that a transverse rotation of the electron mass has to be involved during these interaction states ([49], [8] Chapter 9).

The use of these terms is so generalized however that changing them is likely to cause even more confusion, but the real nature of the states and processes being referred to should be clearly documented in formal reference repositories such as *NIST* [50] and the *CRC Handbook of Chemistry and Physics* [51], for example.

1.11. The simultaneous induction of both energy half-quanta

This new awareness of the simultaneous existence of these two energy half-quanta, mutually perpendicular to each other, that are permanently induced in all charged elementary particles, whether they are in motion or not, and whose amount progressively varies according to the inverse of the distances between each charged particle and all others, now allows establishing at the subatomic level an internal electromagnetic structure of the energy quantum sustaining both the longitudinal momentum increase and the increase of the transverse magnetic field of any accelerating charged elementary particles, which is identical to that suggested by Louis de Broglie in the 1930's for localized electromagnetic photons ([15], [8] Chapter 6), which is in complete agreement with Maxwell's equations, but which is not in contradiction with the manner in which free moving electromagnetic energy has mathematically been successfully dealt with at the macroscopic level from the viewpoint of Maxwell's continuous wave theory.

1.12. Description of Marmet's derivation from Equation (M-1) down to Equation (M-6)

In electromagnetism, the Biot-Savart equation is possibly the easiest equation to confirm experimentally because it only describes the transverse uniform and invariant cylindrical magnetic field generated by a stable continuous electric current flowing in a straight electric wire [19].

Grounding his reasoning on the fact experimentally observed during high energy particles accelerators experiments that the magnetic field of an accelerating electron increases despite the also observed fact that its unit charge remains constant irrespective of its velocity, Marmet succeeded, by theoretically reducing to one electron the current flowing in the wire, to derive Equation (M-23) from the Biot-Savart equation, which allows demonstrating that the transversely measurable relativistic mass increase of an accelerating electron, can only be directly related to its transverse magnetic field increase.

Finally, Equation (M-24) that directly emerges from Equation (M-23), directly establishes that exactly half of the energy making up the invariant rest mass of the electron is also representable as a magnetic field, presumably also transverse by analogy, and would also be in reality an invariant amount of energy that would also be physically oriented transversely:

$$\frac{\mu_0 \left(e^-\right)^2}{8\pi}\frac{1}{r_e} = \frac{M_e}{2} \tag{M-24}$$

This observed characteristic of the intrinsic magnetic field of the rest mass of the electron, among many others that Marmet's discovery allows at long last to correlate in a new mutually self-consistent perspective, will be analyzed further on, as well as the *velocity-dependence* aspect of the accelerating electron increasing transverse magnetic field, as well as the further developments that Equation (M-23) leads to. But let us first address the hurdle presented by Equation (M-7).

He began his derivation by introducing the following form of the Biot-Savart Equation (M-1), in which the cylindrical transverse magnetic field that appears about a current carrying rectilinear metallic wire when a stable electric current is circulating, is represented as being perpendicular to the current direction in the wire, as illustrated in **Figure 1.1** of his paper [29], that is, as being perpendicular to the axis along which current I is graphically represented as flowing:

$$d\vec{B} = \frac{\mu_0 I}{4\pi}\frac{d\vec{s} \times d\vec{u}}{r^2} \tag{M-1}$$

He then redefined current I by quantizing the electron charge to its invariant unit value (e=1.602176462E-19 C), which allowed replacing the general variable charge symbol Q in the standard definition of I by the discrete number of electrons in one ampere:

$$I = \frac{dQ}{dt} = \frac{d(Ne^-)}{dt} \tag{M-2}$$

Since the velocity of electrons in a conductor remains constant if current I remains constant, the time element dt can also be replaced by its traditional definition dx/v:

$$\text{since } v = \frac{dx}{dt} \text{ , then } dt = \frac{dx}{v} \tag{M-3}$$

Replacing now dt of the definition of I as previously established with Equation (M-2) by its equivalent definition established with Equation (M-3), he obtained:

$$I = \frac{d(Ne)}{dt} = \frac{d(Ne^-)v}{dx} \tag{M-4}$$

He then introduced the scalar version of the Biot-Savart equation:

$$dB = \frac{\mu_0 I}{4\pi r^2}\sin(\,\theta\,)dx \tag{M-5}$$

Replacing I in Equation (M-5) by its new definition established with Equation (M-4) also eliminates the implied time factor from Biot-Savart equation, which can be done in context without affecting the value of the magnetic field considered since it remains constant by definition since the current remains constant:

$$dB = \frac{\mu_0 I}{4\pi r^2}\sin(\,\theta\,)dx = \frac{\mu_0}{4\pi r^2}\frac{d(Ne^-)v}{dx}\sin(\,\theta\,)dx = \frac{\mu_0 v}{4\pi r^2}\sin(\,\theta\,)\,d(Ne^-) \tag{M-5a}$$

In summary, Marmet's Equation (M-6) is now presented as follows, now involving a sum of quantized unit charges, represented by factor Ne^-, on top of being disconnected from the time factor, since the magnetic field intensity will remain stable as long as the current remains stable, irrespective of the time elapsed:

$$dB = \frac{\mu_0 v}{4\pi r^2}\sin(\,\theta\,)\,d(Ne^-) \tag{M-6}$$

1.13. The erroneous Equation (M-7) published by mistake

We now reach the equation that seems not to logically emerge from the seamless sequence that led to Equation (M-6) above, which is likely to have caused an undue loss of interest on the part of potentially interested researchers in reading further, which may explain why this article has not attracted more attention up to now:

$$\text{Incorrect Equation (M-7): } \quad dB_i = \frac{N\mu_0 e^- v}{4\pi r^2}\,d(Ne^-) \tag{M-7}$$

It seems also that Paul Marmet did not become aware of this typographical error during the 2 years separating the publication of the article in 2003 from his passing away in 2005, which would explain why he produced no *erratum* note to rectify this misprint, because it is absolutely certain that he derived the correct following form of Equation (M-7), that we will now correctly re-establish, since he used this correct form for the remainder of his derivation:

$$\text{Corrected Equation (M-7): } \quad B_i = \frac{\mu_0 e^- v}{4\pi r^2} \tag{M-7}$$

1.14. Re-establishing the Correct Form of Equation (M-7)

As analyzed by Marmet in his explanatory text between Equations (M-6) and (M-7), two variables of Equation (M-6) will now be reduced to the constant value 1 by structure due to the number of electrons being brought down to a single one in Equation (M-7), in which case the charge distribution and magnetic field distribution become by structure isotropic and spherically centered on the location of this single electron, instead of being respectively conceptually linearly distributed for the charge and in transverse cylindrical distribution perpendicularly to the current direction for the magnetic field, as in the initial Biot-Savart equation. Here is then how the correct Equation (M-7) can be derived from Equation (M-6).

First, the N term in Equation (M-6) will become equal to 1 in Equation (M-7) since only one electron is being considered in the latter equation, so first the term $d(Ne^-)$ will become d(e$^-$), which is the first step in transiting from Equation (M-6) to the correct form of Equation (M-7):

$$dB_i = \frac{\mu_0 v}{4\pi r^2} \sin(\,\theta\,)\,\mathrm{d}(\mathrm{e}^-) \qquad\qquad \text{(M-6a)}$$

Since a single electron is being considered, it becomes impossible to conceptually determine a direction of continuous distribution of the electric charge, because no axis of distribution can now be defined. Consequently the sin (θ) factor that was related to this now non-existent linear distribution also disappears from the equation. So we now have:

$$dB_i = \frac{\mu_0\,v}{4\pi r^2}\,\mathrm{d}(\mathrm{e}^-) \qquad\qquad \text{(M-6b)}$$

Since charge e of the electron is invariant and thus becomes a numerical constant, calculating a derivative for Equation (M-6b) becomes meaningless. Consequently the two occurrences of the derivative operator d simplify out of Equation (M-6b), and we finally end up with the real equation that Marmet obviously meant to be published as Equation (M-7):

$$B_i = \frac{\mu_0 v}{4\pi r^2}\,\mathrm{e}^- \qquad\qquad \text{(M-6c)}$$

then rearranged in the following form that he used as he proceeded with his derivation leading to Equation (M-23):

$$\text{Correct Equation (M-7):} \quad B_i = \frac{\mu_0\, e^-\, v}{4\pi\, r^2} \qquad \text{(M-7)}$$

This is how Marmet succeeded in modifying the Biot-Savart equation from representing the uniform macroscopic cylindrical static magnetic field generated by a stable electric current circulating in a rectilinear metallic wire, to representing the velocity related uniform subatomic theoretically spherical transverse magnetic field increment related to the velocity of a single electron, centered on its moving point-like location as it moves at constant velocity, represented by Equation (M-7).

According to the motion mechanics of electromagnetic energy in the expanded trispatial geometry that will be clarified later, this constant velocity of all electrons in the flow of electrons in a metallic wire is due to each electron being individually *propelled*, so to speak, by an amount of physically existing longitudinally oriented momentum energy ΔK, equal by structure to the transversely oriented energy making up the related transverse magnetic field increment ΔB, both amounts physically existing separately from the energy of which the invariant rest mass of the electron is made.

From this perspective, it turns out that the stable transverse and apparently stationary and uniform magnetic field dB of Biot-Savart Equation (M-1) measurable about the metallic wire simply is the sum of the individual moving transverse magnetic fields of the moving electrons, each electron dragging with it its local magnetic field. Since all electrons in the flow move in the same direction and in close proximity to each other, their individual magnetic fields are all *de facto* forced into mutual parallel magnetic spin alignment due to the inflexible triply orthogonal *electric / magnetic / direction-of-motion-in-space* relationship of electromagnetic energy, to which the energy of every elementary electromagnetic particle is subjected to; which explains why all of the individual magnetic fields of the electrons circulating in the wire are oriented in the same transverse direction about the wire, that results in the establishment of this cylindrical macroscopic transverse magnetic field measurable as being stable at any point along the length of a metallic wire in which a constant current is circulating. This is what the Biot-Savart equation is measuring. And this is why reducing the current to involve a single electron allows defining Equation (M-7) that can account for the velocity related subatomic magnetic field increment of a single electron.

It must be mentioned here that the same forced mutual parallel magnetic spin alignment of unpaired electrons in ferromagnetic materials is also what causes their individual magnetic fields to add up to become measurable at our macroscopic level as a single macroscopic magnetic field, as analyzed in References ([49], [8] Chapter 9) ([52], [8] Chapter 10), and formally described in Reference [51]. This confirms that the establishment of all macroscopically measurable magnetic fields, be they dynamic or static, can only be due to the same subatomic process, which is the forced parallel alignment of the magnetic spins of the energy of the elementary electromagnetic quanta involved.

We will see further on how Marmet's Equation (M-7) was generalized to calculate the magnetic field increment of any localized electromagnetic quantum, leading then to generalized forms allowing to calculate the velocity of any charged elementary massive electromagnetic particle by combining the intrinsic invariant magnetic field B of its rest mass with the varying magnetic field ΔB of this energy of motion induced in electrically charged massive particles by the Coulomb interaction.

The remainder of Marmet's derivation up to his determining conclusion represented by equivalence (M-26) is available in his paper [29], and is also analyzed in detail at the beginning of Reference ([10], See also Chapter 2):

$$\textbf{relativistic mass} \equiv \textbf{magnetic mass} \qquad\qquad \text{(M-26)}$$

1.15. The Implications of Marmet's discovery

The first major consequence of the establishment of the Equation (M-23) is the establishment of electromagnetic equations that allow the calculation of the relativistic velocities of charged and massive elementary particles without any need to use the Lorentz γ factor.

1.16. Calculating relativistic velocities without the Lorentz γ factor

Considering Equation (M-23) again, since c constitutes *an asymptotic velocity limit* that the electron cannot physically reach, then as v tends towards c, $M_e/2$

seems to tend towards an asymptotic transverse mass increment limit equal to 4.55469094E-31 kg, corresponding to its transverse magnetic field increment, that apparently seems, at first glance, impossible to increase further, but we will see further on that this is not the case:

$$\frac{\mu_0\left(e^-\right)^2}{8\pi}\frac{1}{r_e}\frac{v^2}{c^2}=\frac{M_e}{2}\frac{v^2}{c^2} \tag{M-23}$$

At this stage of the analysis, Equation (M-23) can thus be formulated as follows to represent the electron transverse *relativistic-mass/magnetic-field* increment:

$$\Delta m_{m(v\to c)}=\frac{\mu_0 e^2}{8\pi r_e}\frac{v^2}{c^2}=\frac{m_e}{2}\frac{v^2}{c^2} \tag{1.1}$$

On the other hand, when v tends towards zero in Equation (M-23), its transverse magnetic field increment also tends towards zero. And when this velocity approaches zero, the ratio v^2/c^2 reveals that the amount of energy of the transverse increment of the magnetic field becomes negligible and that this ratio can then be removed from the equation, which still leaves part of the invariant rest mass an electron as being represented as a magnetic field, apparently finally revealing that exactly half of the energy making up the invariant rest mass of the electron would also be the source of its intrinsic invariant magnetic field, as represented by Equation (M-24), which is a conclusion that will be confirmed further on by the establishment of the Maxwell equation compliant LC Equation (1.30) that reveals the actual inner electromagnetic structure of the electron rest mass energy, that was previously established in the trispatial geometry in relation with de Broglie's hypothesis (**Figure 1.3**):

$$M_{e\_magnetic\,(v\to 0)}=\frac{\mu_0\left(e^-\right)^2}{8\pi}\frac{1}{r_e}\frac{v^2}{c^2}=\frac{\mu_0\left(e^-\right)^2}{8\pi}\frac{1}{r_e}=\frac{M_e}{2} \tag{M-24}$$

Equation (M-7), on the other hand, can be formulated as follows to represent the corresponding transverse magnetic field increment that represents the same amount of increasing energy measurable as the transverse mass increment represented by Equation (1.1) which adds to that of the invariant magnetic field of the electron's rest mass, calculable with Equation (M-24):

$$\Delta B_{(v\to c)}=\frac{\mu_0\,e\,v}{4\pi\,r^2} \tag{1.2}$$

As a first step in confirming that Equations (1.1) and (1.2), both are representations of the same amount of transversely oriented energy in relation with the direction of motion of the accelerating electron, let us first resolve Equation (1.1) for a well known relativistic velocity, that is, velocity 2187647.561 m/s related to the Bohr ground orbit momentum energy in his theory about the hydrogen atom (2.179784832E-18 j), which also happens to be the real mean momentum energy given by the wave function of Quantum Mechanics for the electron ground state orbital of the hydrogen atom. This velocity will immediately confirm that Equation (1.1) provides the correct related relativistic mass increment:

$$\Delta m_m = \frac{\mu_0 e^2 v^2}{8\pi\, r_e c^2} = \frac{\mu_0 e^2 (2187647.561)^2}{8\pi\, r_e c^2} = 2.425337715E - 35\,\text{kg} \qquad (1.3)$$

By means of Equation (1.2), which is, let us remember, Marmet's Equation (M-7), we must now calculate the increase in the transverse magnetic field associated with this same relativistic velocity of the electron. For this purpose, we must define the value of the second variable in Equation (1.2), that is, the value of r; and it cannot outright be assumed that it will have the same value r_e of Equation (1.1), which is a constant known as the *classical electron radius*, used in this equation in relation with the electron rest mass.

In the case of Equation (1.1), that is, Marmet's Equation (M-23) combining an electromagnetic definition of the electron mass with its classical/relativistic mechanics definition, a close examination shows that the relativistic-mass/magnetic-field increment can only synchronously increase with the velocity ratio v^2/c^2, c being invariant and v ranging from zero to asymptotically close to c, which, as previously mentioned, seems to reveal that the theoretical maximum possible increment of transverse relativistic-mass/magnetic-field of a free moving electron seems not to really tend towards infinity as traditionally anticipated, but would rather tend to become asymptotically close to a value equal to half the invariant rest mass of the electron ($\Delta m_m = m_e/2 = 4.55469094E-31$ kg, corresponding to the induced transverse energy half-quantum of 4.09355207E-14 j).

Let us remember at this point that the Marmet Equation (M-23) defines the relativistic-mass/magnetic-field increment as being dependent strictly on the value of the invariant half of the rest mass energy of the electron that defines its intrinsic invariant magnetic field. But a conversion to electromagnetic form of the classical Newton kinetic energy equation $K = mv^2/2$ completed by its correction to incorporate the transverse magnetic energy identified by Marmet and that was missing in Newton's Equation ([42], [8] Chapter 5), ultimately shows that as the transverse magnetic field increases, any further increase of this transverse relativistic-mass/magnetic field increment does not depend uniquely on half the energy of the electron rest mass, as non-relativistic Equation (M-23) suggests, but depends in reality on the total amount of momentarily accumulated transverse energy, that is, on the sum of the energy making up the mass of the intrinsic magnetic field of the electron $m_e c^2/2$ plus the energy of the momentarily accumulated transverse mass increment $\Delta m_m c^2$.

This means that the transversely measurable relativistic mass of an accelerating electron $m_{relativistic}$ is always equal to $m_0+\Delta m_m$, which allowed establishing that this sum is always equal to the invariant rest mass of the electron multiplied by the well known gamma factor γm_o that was established more than one century ago ([42], [8] Chapter 5). This is what allows calculating the whole range of relativistic velocities of the electron without using the gamma factor (known as the Lorentz factor).

For example, any relativistic velocity of an electron can be calculated with the following equation derived in Reference ([42], [8] Chapter 5), by setting E to 8.18710414E-14 j, that is, the energy of the invariant rest mass of the electron, and setting K to the sum of energy of the transverse relativistic-mass/magnetic-field increment $\Delta m_m c^2$ plus the related momentum energy ΔK that we now know is always equal by structure to $\Delta m_m c^2$, that is, $K= \Delta K+ \Delta m_m c^2$:

$$v = c\frac{\sqrt{4E\cdot K+K^2}}{2E+K} \qquad (1.4)$$

This equation can also be converted to a form making use of the wavelengths of the energies involved ([42], [8] Chapter 5), allowing the very same calculation of the whole range of relativistic velocities of the electron strictly from the wavelengths of the energies involved:

$$v = c\frac{\sqrt{4\lambda\cdot \lambda_C + \lambda_C^2}}{2\lambda+\lambda_C} \qquad (1.5)$$

From this equation, the gamma factor was directly derived as analyzed in Reference ([42], [8] Chapter 5), thus bringing the proof of the validity of Marmet's derivation that allowed the elaboration of these equations.

1.17. A cause more fundamental than velocity for the induction of momentum and transverse magnetic field energy

Let us now return to the correlations that must be made between Equations (1.1) and (1.2). We observe in the electromagnetic definition of mass of Equation (1.1), that it is the *classical radius* of the electron r_e that connects this equation to the concept of mass. In the case of Equation (1.2), which emerges strictly from electromagnetism, it is also clear that the transverse magnetic field can only increase according to the same velocities ratio, because Marmet's demonstration clearly reveals that the energy half-quantum represented by mass increment Δm_m in Equation (1.1) is the same transversely oriented energy half-quantum which is also described by the transverse magnetic field increment ΔB; but the value that r must have in Equation (1.2) for the energy corresponding to the increase of ΔB to coherently vary from zero to this asymptotic limit made up of the sum of the energy of the classical half-quantum of the electron's rest mass of 4.09355207E-14 j plus the momentarily accumulated energy of ΔB, is not clearly established. To understand what this value should be, we must now understand the relation between r_e used in Equation (1.1) and the mass of the electron, or more precisely its relation with the energy constituting the invariant rest mass of the electron.

In a paper published in 2007 in the same Kazan State University International IFNA-ANS Journal ([30], [8] Chapter 4), that describes a first wave of conclusions emerging from Marmet's discovery, it was conclusively established that r_e is in reality the lower limit of spherical integration of the energy making up the invariant rest mass of the electron ($E=m_ec^2$=8.18710414E-14 j), and that r_e turns out to be in reality the *transverse amplitude of electromagnetic oscillation* of the energy making up the measurable rest mass of the electron, which is obtained by multiplying the electron Compton wavelength by the fine structure constant α and dividing them by 2π, as determined in Reference ([31], [8] Chapter 11):

$$r_e = \frac{\lambda_C\,\alpha}{2\pi} = 2.817940285\text{E}-15\,\text{m} \tag{1.6}$$

Consequently, and by similarity, the value of r that must be used in Equation (1.2) should thus also be that of the *transverse amplitude of electromagnetic oscillation* of the energy induced at the Bohr radius (4.359743805E-18 j), whose *longitudinal electromagnetic wavelength* would be (λ=4.556335256E-8 m) if it was moving at velocity c, but that must already be multiplied by α to reach the value of the *longitudinal de Broglie wavelength* corresponding, for this energy, to the length of the Bohr orbit, whose radius is (r_B=5.291772083E-11 m), keeping in mind that this radius remains valid in Quantum Mechanics since it is exactly equal to the mean axial resonance distance of the electron within the volume defined by Schrödinger's wave equation for the electron captive in the hydrogen ground state orbital ([10], See also Chapter 2):

$$r_B = \alpha\, r = \frac{\alpha\, \lambda}{2\pi} = \frac{\lambda_B}{2\pi} = 5.291772083\,\text{E}-11\,\text{m} \tag{1.7}$$

By similarity to the method used with Equation (1.6) to define the *transverse amplitude of electromagnetic oscillation* of the electron rest mass energy by multiplying the *longitudinal electromagnetic wavelength* λ_C of its energy by α, there is thus need to multiply also the *longitudinal de Broglie wavelength* λ_B defined in Equation (1.7) for the energy induced at the Bohr radius r_B again by α to finally reach the *transverse* value αr_B of the *transverse amplitude of the oscillation* of the electromagnetic energy induced at the Bohr radius (αr_B = 3.861592641E-13 m), which now makes it possible to establish the intensity of the transverse magnetic field increment $\mathit{\Delta B}$ which becomes measurable by being added for the velocity considered to the invariant transverse magnetic field of the rest mass of the electron. Let's now calculate the magnetic field corresponding to relativistic velocity 2187647.561 m/s and to this value of $r=\alpha r_B$ with Equation (1.2):

$$\mathit{\Delta \mathbf{B}} = \frac{\mu_0\, e v}{4\pi \left(\alpha\, r_B\right)^2} = \frac{\mu_0\, e\left(2187647.561\right)}{4\pi \left(\alpha \times 5.291772083E-11\right)^2} = 235047.0405\,\text{T} \tag{1.8}$$

It is interesting to note by the way that r_e, as calculated with Equation (1.6), is only distant from an additional multiplication by α from the value of αr_B, as established in Reference ([53], [8]), which suggests a possible axial resonances sequence establishing a sequence of stable stationary action electromagnetic states whose unit of axial progression seems to be the fine structure constant α, as put in perspective in the same reference.

To confirm the validity of the value obtained with Equation (1.6), which is also measurable as a transverse magnetic mass increment $\mathit{\Delta m_m}$ with Equation (1.3), let's calculate it with Equation (1.9) which is the generalized version of Marmet's Equation (M-7) that was established in the 2007 article ([30], [8] Chapter 4). Unlike Equation (M-7), it can be observed that this generalized form does not require using the velocity of the particle to obtain the intensity of its transverse magnetic field increment.

Only the *longitudinal electromagnetic wavelength* of the total carrier energy of the electron is required, that is, the energy of its momentum plus the transverse energy representable either as a magnetic mass increment Δm_m or as a magnetic field increment $\boldsymbol{\Delta B}$. Since the total energy induced at the Bohr orbit is (E=4.359743805E-18 j), its *longitudinal electromagnetic wavelength* is thus ($\lambda = hc / E = $ 4.556335256E-8 m), and we obtain with this generalized equation the same value as with Equation (1.8):

$$\Delta \mathbf{B} = \frac{\mu_0 \, \pi e c}{\alpha^3 \lambda^2} = \frac{\mu_0 \, \pi e c}{\alpha^3 \left(4.556335256 E - 8 \right)^2} = 235051.7346 \, \mathrm{T} \tag{1.9}$$

We thus observe that without any need to imply any velocity, generalized Equation (1.9) provides in Tesla the very same transverse magnetic field increment energy density as the initial Marmet Equation (M-7) originally derived from the Biot-Savart equation, in which the intensity of the transverse magnetic field increment *seems to depend* on the velocity of the particle, since that in the Biot-Savart equation from which it was derived, the intensity of the increment of the magnetic field varies strictly according to the velocity of the electrons circulating in the wire.

The fundamental question that now comes to mind is the following, when considering Equation (1.9): "*How come that it is possible, to calculate the correct intensity of the 'supposedly' velocity dependent variable transverse magnetic field increment of a moving electron, without this velocity being used to calculate it?*".

1.18. Momentum and transverse magnetic field energy increase without velocity increase

This difference between Equation (M-7), that requires the use of a velocity to calculate the related intensity of the transverse magnetic field increment of the moving electron, and its generalized version used to solve Equation (1.9), that does not require this velocity, draws attention to a cause more fundamental than motion to explain the induction of energy in the electron even when no velocity is involved.

It is a long established fact in classical mechanics, from direct observation, that the kinetic energy traditionally named the *energy momentum* of a macroscopic mass in motion depends strictly on its velocity, and that this energy is considered to be the only motion related energy that exists in excess of the energy making up the rest mass of a massive body. The amount of energy of this momentum of an accelerating macroscopic mass is consequently defined in classical mechanics as increasing rectilinearly, potentially without limit, only due to its velocity increase, itself also deemed to possibly be increasable without limit.

This definition of the increasing kinetic momentum of an accelerating macroscopic mass is also admitted in Special Relativity with this difference that the momentum energy is defined as increasing according to a non-rectilinear curve that we know is correct, also theoretically without limit, as the velocity increases, but that this potentially infinite value would be reached before the velocity of light is reached, this velocity being defined as an unreachable asymptotic velocity limit deemed impossible to be reached by massive bodies. Confirmation of the accuracy of equation $K=m_oc^2(\gamma-1)$ from Special Relativity was never obtained, however, by means of macroscopic masses in motion, since we do not have the technology required to accelerate macroscopic masses to relativistic velocities, but rather using the subatomic mass of the electron, with which the accuracy of this equation was confirmed by Kaufman's first experiments [36].

As put in perspective earlier in this chapter, it must be understood that as the theory of Special Relativity was being developed, the fact that the invariant rest mass of the electron m_0=9.10938188E-31 kg is also the seat of its invariant unit electric charge e=1.602176462E-19 C was not yet understood as meaning that the Coulomb interaction, that induces the energy of the momentum and of the transverse magnetic field in all electrically charged elementary particles such as electrons strictly as a function of the inverse of the distance between them, and this, even if this distance does not vary, induces it *de facto* at the same time with respect to the rest mass of these charged and massive particles, since the charge and the mass of the electron are two characteristics of the same elementary particle.

Considering that the mass of all macroscopic bodies can only be the sum of the subatomic masses of the massive elementary particles of which it is made, how then can this be reconciled the fact that no increase in the magnetic field of any moving macroscopic masses seems to ever have been measured, since this increase is easily measurable for an accelerating electron, as abundantly demonstrated experimentally since Kaufman's first experiments [36], which also provides experimental confirmation of the non-rectilinear growth of the momentum energy of an accelerating electron towards this theoretical infinite quantity that the asymptotic limit imposed by the speed limit of light suggests?

Indeed, such relativistic-mass/magnetic-field increments of macroscopic masses may well have been detected all the same for much lower velocities than those typical of electrons, but having been seen as "anomalies" or rationalized otherwise, instead of having been recognized as such, due to the fact that the Special Relativity theory on which all analysis of relativistic effects are grounded does not recognize its existence, as previously put in perspective, and as we will now observe from experimental data.

1.19. The "anomalous" trajectories of the Pioneer 10 and 11 space probes

As previously mentioned, it must be realized here that it has never been possible to accelerate macroscopic masses to velocities comparable to those to which electrons are typically accelerated to at the subatomic level, that were sufficient to confirm the non rectilinear increase of their momentum energy accounted for by SR, and that are also sufficient to confirm the simultaneous increase in transverse magnetic field energy which is not accounted for by SR.

The highest velocities reached by macroscopic projectiles launched into space have currently been reached by the Pioneer 10 and Pioneer 11 space probes, with respective approximate masses made available by NASA of 258 kg and 258.5 kg, as measured before liftoff. Their velocities varied greatly throughout their trajectories, with peaks of 132,000 km/h (36667 m/s) for Pioneer 10, which is its peak velocity during its final acceleration by gravitational slingshot using Jupiter, and 175,000 km/h (48611 m/s) for Pioneer 11, which is its peak velocity during its final acceleration by gravitational slingshot using Saturn.

We will analyze here more specifically the escape velocities of the two probes. The reader can make the calculations for the peak velocities mentioned above, that would reveal the increase in mass that can explain the so-called *anomalous* velocity peaks [48] observed during these acceleration phases of the two probes, as well as during the similar phases of all other space probes subjected to gravitational slingshot acceleration, and that leave the entire astrophysical community perplexed and without explanation, because the SR theory that currently serves as the basis for any analysis of these trajectories is unable to account for them.

These two space probes respectively reached escape velocities of 51682 km/h (14356 m/s) and 51800 km/h (14389 m/s), which are velocities 150 times lower than the theoretical velocity of 2187647.561 m/s of the electron in the theoretical Bohr's ground state orbit, for which the increment of its transverse magnetic field is just beginning to be experimentally measurable (See Equation (1.3)).

What is remarkable about the trajectories of these space probes, as well as about those of all other space probes launched throughout the solar system, is that an unexplained systematic anomaly was noted. Without exception, they behave as if they were slightly more massive than their masses as measured before liftoff, showing a systematic negative acceleration of about 8E-6 m/s towards the Sun [46] [47] [48].

But as Rainer W. Kühne mentions in a note published in 1998, the extensive publicity given to these two cases leaves the general impression that this problem concerns only space probes launched by man [54], but it is well known in the astrophysics community that the trajectories of planets Uranus, Neptune and Pluto also show similar systematic anomalies, as well as many comets already studied in 1998, such as Halley, Encke, Giacobini-Zinner and Borelli, whose trajectories undergo a systematic deviation of unknown origin.

Given the understanding now provided by Marmet's discovery, even with the relatively low velocities of the Pioneer 10 and 11 space probes with respect to the typically relativistic velocities of the electron, it becomes easy to calculate this transverse energy increment of the relativistic-mass/magnetic-field that increases the transverse inertia of these two space probes, because we know now for certain by structure that the amount of transverse energy induced at the same time as their momentum is always equal to the latter. The characteristics of the two probes being almost identical, we will use the parameters of Pioneer 10 to analyze this situation.

So, with m=258 kg and v=14356 m/s, we first obtain the momentum energy of Pioneer 10 for this escape velocity:

$$\Delta K = mc^2 \left(\frac{c}{\sqrt{c^2 \text{-} v^2}} - 1 \right) = 2.658722735\text{E}10\,\text{j} \tag{1.10}$$

Given that the energy of Δm_m is equal by structure to ΔK, we then obtain for Pioneer 10 a transverse increment of relativistic-mass/magnetic-field of:

$$\Delta m_m = \frac{\Delta K}{c^2} = 2.958228\text{E} - 7\,\text{kg} \tag{1.11}$$

Such a slight transverse inertia increase seems insufficient at first glance to explain on its own the systematic negative acceleration of about 8E-6 m/s towards the Sun of these space probes launched on escape trajectories from the solar system, but the proposal becomes much more likely if we add to it the adiabatic increase of the rest mass of each probe due to the initial phase of their trajectory away from the incommensurably larger mass of the Earth, that is, an adiabatic rest mass increase that was easily observed during the famous Hafele and Keating experiment [55] when atomic clocks were raised just 10 km from the Earth's surface, but was misinterpreted as confirming a variation in the rate of time flow ([45], [8] Chapter 16), here again only in light of the theory of General Relativity, that doesn't take into account the involvement of the Coulomb interaction, nor the fact that all rest masses are exclusively made of electrically charged particles. This adiabatic increase in rest masses will be put in correct electromagnetic perspective in Section 1.27.

1.20. Maximum Intensity of the transverse magnetic field increment

Coming back now to the comparison between generalized Equation (1.9) and Equation (1.8), which is actually Marmet's Equation (M-7), we observe that Equation (1.9) provides the same magnetic field energy density in Tesla as Equation (M-7), but requires only one variable and not two like Equation (M-7), that is, only the *longitudinal electromagnetic wavelength* of the energy quantum involved, without any need to relate this energy with the electron velocity.

This is what makes this magnetic field equation general and appropriate for calculating the intrinsic magnetic field of any elementary electromagnetic particle, whether it is moving or not. For example, the invariant intrinsic magnetic B_e field of the electron, that accounts for half of its invariant rest mass energy, can be calculated as follows, using the electron Compton wavelength, also involving the fine structure constant that establishes the amplitude of this energy's transverse electromagnetic oscillation:

$$\mathbf{B}_e = \frac{\mu_0 \, \pi \, e \, c}{\alpha^3 \lambda_C^2} = \frac{\mu_0 \, \pi \, e \, c}{\alpha^3 \left(2.426310215 E\text{-}12\right)^2} = 8.289000221 E13 \, \text{T} \qquad (1.12)$$

Of course, this figure remains mostly meaningless without a solid confirmation that it really represents a physically existing *quantity*, that is, a confirmation that could be obtained by showing that relativistic velocity

v=2187647.561 m/s, related to the magnetic field energy density calculated with Equation (1.9), for example, can really be calculated by providing only the electromagnetic wavelength of the related energy as the only variable in an equation otherwise involving only fundamental physical constants.

Such a confirmation can indeed be provided by means of the following equation, well known in high energy accelerator circles, that allows calculating the straight line relativistic velocity of an electron being accelerated by external equal intensities electric and magnetic fields:

$$v = \frac{\mathbf{E}}{\mathbf{B}} \tag{1.13}$$

The proper value for the required composite $\mathbf{B}$ field is established by simple addition of Equations (1.9) and (1.12), as analyzed in Reference ([30], [8] Chapter 4), here calculated with the longitudinal wavelength of the energy induced at the Bohr ground state radius (λ=4.556335256E-8 m), to account for the required ΔB field increment, and the electron longitudinal Compton wavelength (λ_C=2.426310215E-12 m) to account for the invariant internal $\mathbf{B_e}$ field of the rest mass of the electron:

$$\mathbf{B} = \mathbf{B_e} + \Delta\mathbf{B} = \frac{\mu_0\,\pi\,e\,c}{\alpha^3\lambda_C^{\,2}} + \frac{\mu_0\,\pi\,e\,c}{\alpha^3\lambda^2} = \frac{\mu_0\,\pi\,e\,c}{\alpha^3}\,\frac{\left(\lambda^2 + \lambda_C^{\,2}\right)}{\lambda^2\lambda_C^{\,2}} = 8.28900024\,6\text{E}13\ \text{T} \tag{1.14}$$

Resolving Equation (1.13) also requires of course the establishment of the definition of the composite E field that must be set in equilibrium with this composite $\mathbf{B}$ field. The related general E field equation was also established in Reference ([30], [8] Chapter 4), thanks to reformulation of the Coulomb equation established in the same article, a reformulation that was analyzed in depth in Reference ([10], See also Chapter 2) and that allows calculating the amount of transverse oscillating energy that generates and maintains the corresponding oscillating magnetic field in elementary electromagnetic particles, whatever state of least action motion or of electromagnetic equilibrium stationary action they may be involved into within atomic structures:

$$E = \int_{a_0}^{\infty} \frac{1}{4\pi\,\varepsilon_o}\,\frac{e^2}{\left(\alpha\,\lambda/2\pi\right)^2} \cdot dr = 0 - \frac{1}{4\pi\,\varepsilon_o}\,\frac{e^2\,2\pi}{\alpha\,\lambda} = \frac{e^2}{2\,\varepsilon_o\alpha\lambda} \tag{1.15}$$

This particular form of the Coulomb equation indeed allows calculating the energy of any electromagnetic quantum uniquely from its wavelength, without any need to use the Planck constant:

$$E = hf = \frac{e^2}{2\varepsilon_o \alpha \lambda} \tag{1.16}$$

As mentioned already in Subsection 1.7.3, this form of the Coulomb equation also allowed unifying all classical force equations in Reference ([44], [8] Chapter 7) by showing that the *F=ma* fundamental acceleration equation can be derived from all of them, which actually proves that the Coulomb interaction is the common denominator of all classical force equations.

The general *E* field equation corresponding to the general *B* field Equation (1.9) was thus established in Reference ([30], [8] Chapter 4) as follows, here resolved using the longitudinal wavelength of the energy induced at the Bohr ground state (λ=4.556335256E-8 m), to harmonize with the *ΔB* field value obtained with Equation (1.9):

$$\Delta E = \frac{\pi e}{\varepsilon_0 \alpha^3 \lambda^2} = 7.046673727E13 \text{ N/C} \tag{1.17}$$

Consequently, the invariant E_e field related to the other half of the energy making up the invariant rest mass of the electron can be established with the electron longitudinal Compton wavelength as follows:

$$E_e = \frac{\pi e}{\varepsilon_0 \alpha^3 \lambda_C^2} = 6.029331754E10 \text{ N/C} \tag{1.18}$$

But, contrary to the composite magnetic *B* field of Equation (1.14) that must be used to calculate the relativistic velocity of the electron with Equation (1.13), and which is obtained from the simple addition of the electron's intrinsic invariant B_e field and of the *ΔB* field of its velocity related magnetic field increment, the corresponding composite *E* field involving the E_e-field and the *ΔE* field of Equations (1.17) and (1.18), cannot be obtained in this simple manner, because the combination of the electric dipole that induces the accompanying *ΔE* field, which is oriented perpendicularly within Y-space with respect to the monopolar E_e-field of the electron's rest mass, involves a vector cross-product to be combined with the E_e field, as clarified in Reference ([31], [8] Chapter 11). As established in Reference ([30], [8] Chapter 4), this composite *E* field, also involving here both the wavelength of the Bohr ground state energy (λ=4.556335256E-8 m) and the electron Compton wavelength (λ_C=2.426310215E-12 m), will have the following value:

$$E = \frac{\pi e}{\varepsilon_0 \alpha^3} \frac{\left(\lambda^2 + \lambda_c^2\right)\sqrt{\lambda_C(4\lambda + \lambda_C)}}{\lambda^2 \lambda_C^2 \ (2\lambda + \lambda_C)} = 1.813341121 \text{ E20 N/C} \tag{1.19}$$

By means of Equation (1.13) the well known and exact relativistic velocity of an electron whose magnetic field is increased by an amount of ΔB will then be obtained as follows if it is not impeded by the local electromagnetic equilibrium state, by means of the values calculated with Equations (1.14) and 1.19):

$$v = \frac{E}{B} = \frac{1.81334112 \text{1E20}}{8.289000246\text{E}13} = 2,187,647.566\,\text{m/s} \tag{1.20}$$

Calculating with Equation (1.9) for the ΔB field and with Equation (1.17) for the ΔE field with any longitudinal wavelength of the electron carrying energy will mathematically show that by combining them with the B_e and E_e fields that account for the energy of the invariant rest mass of the electron obtained with Equations (1.12) and (1.18) to ultimately resolve Equation (1.13), that the whole range of all relativistic velocities up to the asymptotic limit of the speed of light can be obtained in the case of any elementary massive particle such as the electron, and this, for a very mechanical reason which is clearly explained in Reference ([42], [8] Chapter 5).

1.21. Separation of the electron carrying energy from the energy of its rest mass

As analyzed in Reference ([30], [8] Chapter 4), the most significant progress resulting from Marmet's derivation was the new possibility of clearly separating the invariant amount of energy constituting the electron's rest mass from the variable adiabatic energy supporting its motion and its transverse relativistic-mass/magnetic-field increment. After analysis, this variable adiabatic carrying energy of the electron turned out to have the same internal electromagnetic structure that Louis de Broglie proposed for the double-particle electromagnetic photon in the 1930's ([15], [8] Chapter 6) [27] [53], as mathematically described with Equation (1.21), and graphically symbolized with **Figure 1.4**, in accordance with Maxwell's interpretation, according to which the electromagnetic component of the energy of a localized photon has to be oriented transversely with respect to its momentum energy, and is captive in a standing oscillation motion causing it to cyclically transit between a state corresponding to its electric field and a state corresponding to its magnetic field.

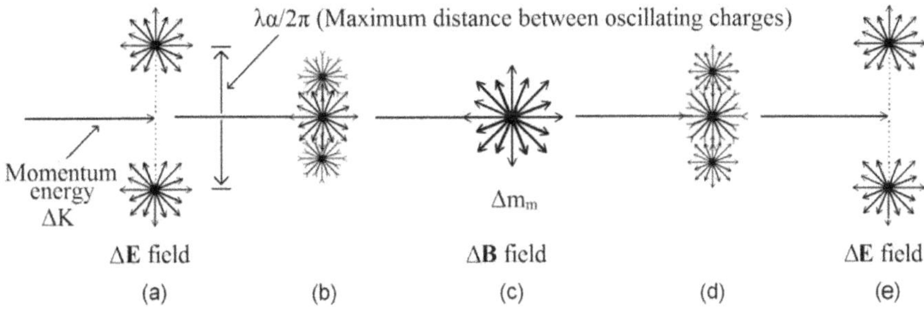

Figure 1.4. Representation of the transverse oscillation cycle of the electromagnetic half-quantum of the electron carrier-photon and of its unidirectional momentum energy half-quantum that propels this first half-quantum on top of also propelling the complete quantum of the electron invariant rest mass energy (the latter not illustrated).

This is what justified coining the term "*carrier-photon*" to name the carrying energy of the electron or that of any other elementary charged particle in articles describing the various consequences of integrating Marmet's discovery into electromagnetic theory on the one hand, and into classical/relativistic mechanics on the other, with the consequence that their equations can now be derived from each other ([10], See also Chapter 2).

The LC equation for the de Broglie double particle photon thus established in the only manner possible in the trispatial geometry proposed at the event Congress-2000 [28], and as formally published in Reference ([15], [8] Chapter 6) in complete accordance with Maxwell's equations, already made it possible to calculate from an electromagnetic photon's wavelength, the maximum intrinsic magnetic field energy of a photon structured according to Maxwell's initial interpretation that both fields induce each other, as established in Reference ([53], [8]):

$$E = \frac{hc}{2\lambda} + \left[\frac{e^2}{2C_\lambda} \cos^2(\omega t) + \frac{L_\lambda \, i_\lambda^2}{2} \sin^2(\omega t) \right] \qquad (1.21)$$

where

$$E_{\mathbf{E(max)}} = \frac{e^2}{2C_\lambda} \qquad \text{and} \qquad E_{\mathbf{B(max)}} = \frac{L_\lambda \, i_\lambda^2}{2} \qquad (1.22)$$

and

$$C_\lambda = 2\varepsilon_0 \alpha\lambda \qquad L_\lambda = \frac{\mu_0 \alpha\lambda}{8\pi^2} \qquad i_\lambda = \frac{2\pi\, ec}{\alpha\lambda} \tag{1.23}$$

Marmet's derivation, on its part, made it possible to establish in Reference ([30], [8] Chapter 4) the generalized electric and magnetic field equations previously mentioned, that directly match the representations of their energy in the form of capacitance and inductance as illustrated by Equations (1.22):

$$\mathbf{E} = \frac{\pi e}{\varepsilon_0 \alpha^3 \lambda^2} \qquad \mathbf{B} = \frac{\mu_0 \pi ec}{\alpha^3 \lambda^2} \tag{1.24}$$

and to also establish *the theoretical stationary isotropic volume* corresponding to the maximum energy density of each of these two mutually inducing fields:

$$V = \frac{\alpha^5}{2\pi^2}\,\lambda^3 \tag{1.25}$$

which made it possible to redefine in Reference ([15], [8] Chapter 6) the LC equation initially developed in Reference ([30], [8] Chapter 4) in a form making use of the more familiar E and B fields definitions, which confirmed that the localized electromagnetic photon as de Broglie conceived it and the electron carrying energy actually have the same internal electromagnetic structure, *i.e.* one half oriented longitudinally, supporting its momentum, and the other half oriented transversely, defining its E and B fields inducing each other, this transversely oriented energy half being propelled in space by the unidirectional energy of its momentum:

$$E = \left(\frac{hc}{2\lambda}\right) + \left[2\left(\frac{\varepsilon_0 \mathbf{E}^2}{4}\right)\cos^2(\omega t) + \left(\frac{\mathbf{B}^2}{2\mu_0}\right)\sin^2(\omega t)\right]V \tag{1.26}$$

1.22. Conversion of electromagnetic energy into charged and massive elementary particles

We have the experimental proof since Carl David Anderson's experiments in 1933 [23] that any electromagnetic photon of energy 1.022 MeV or more, generated as a by-product of cosmic radiation, will destabilize when grazing a massive atomic nucleus, and will convert into a pair of massive elementary particles, which are one electron and one positron, whose equal rest masses of 0.511 MeV/c^2 are each made of 0.511 MeV of the destabilizing photon energy. Any energy in excess of this specific amount of 1.022 MeV that the photon had before conversion is then expressed as longitudinal momentum energy and related transverse electromagnetic energy equally shared between both elementary massive particles, which causes them to move away from each other with a velocity corresponding to this excess momentum energy ([31], [8] Chapter 11).

The following equation describes how the energy of the incident photon is distributed between the two charged and massive particles generated, by associating the Coulomb equation with the rest mass equation of classical/relativistic mechanics ([10], See also Chapter 2). It should be noted in passing that the opposite charges of the electron and the positron are meaningless in classical/relativist mechanics, and that considered according to only their mass characteristic, they are identical, which makes it possible to build the equation in the following way:

$$E_{\left(\frac{1}{\lambda_1}\geq\frac{1}{2\lambda_C}\right)} = \frac{e^2}{2\varepsilon_o\alpha}\frac{1}{\lambda_1} = 2\left(\Delta K + \Delta m_m c^2 + m_0 c^2\right) \tag{1.27}$$

where

$$\left(\Delta K + \Delta m_m c^2\right) = \frac{e^2}{2\varepsilon_o\alpha}\frac{1}{\lambda_2} \quad \text{in which} \quad \frac{1}{\lambda_2} = \frac{1}{2}\left(\frac{1}{\lambda_1} - \frac{1}{2\lambda_C}\right) \tag{1.28}$$

In Equation (1.27), m_0 represents the identical individual rest masses of the electron and positron, and λ_1 is the electromagnetic wavelength of the incident photon being destabilized, while in Equation (1.28), λ_2 is the wavelength of the residual energy in excess of the energy of 1.022 MeV that just converted into the invariant rest masses of the two particles, after separation of this residual energy in equal parts between the two now separate particles.

More interesting yet, an experiment carried out in 1997 at the Stanford Linear Accelerator (SLAC), *i.e.* experiment #e144, confirmed that by converging two sufficiently concentrated electromagnetic photon beams towards a single point in space, one beam involving electromagnetic photons exceeding the 1.022 MeV threshold, massive electron/positron pairs were generated without any massive atomic nuclei being close by [24]. This last experiment opens up an entirely new perspective on the possible origin of the universe, as analyzed in Reference ([56], [8] Chapter 17).

The interest of the trispatial geometry developed from the expansion in the form of 3 perpendicular vector spaces emerging from the three-way orthogonal relationship of the vector product of the fundamental E and B vectors of electromagnetism (**Figure 1.3**), is that the more complete vector harnessing now applicable to Equation (1.26) in the following way, as analyzed in Reference ([15], [8] Chapter 6), allowed establishing for the first time in Reference ([31], [8] Chapter 11) a clear mechanics of conversion of the energy of an electromagnetic photon of 1.022 MeV or more, which is only partially oriented perpendicularly to the energy of its momentum, into the invariant amount of energy completely oriented transversely constituting the internal structure of the individual rest masses m_0 of the electron and positron represented in Equation (1.27), i. e. the following equation:

$$E\,\vec{I}\,\vec{i} = \left(\frac{hc}{2\lambda}\right)_X \vec{I}\,\vec{i} + \left[2\left(\frac{\varepsilon_0\mathbf{E}^2}{4}\right)_Y (\,\vec{J}\,\vec{j}\,,\vec{J}\,\overleftarrow{j}\,)\cos^2(\omega t) + \left(\frac{\mathbf{B}^2}{2\mu_0}\right)_Z \overleftrightarrow{K}\,\sin^2(\omega t) \right] V \tag{1.29}$$

converting into the following two equations to represent the internal electromagnetic structure of the rest masses of the electron and positron:

$$m_{e_0}\vec{0} = \frac{V_m}{c^2}\left\{ \left[\frac{\varepsilon_0\mathbf{E}^2}{2}\right]_Y \vec{J}\,\vec{i} + \left[2\left(\frac{\varepsilon_0\mathbf{V}^2}{4}\right)_X (\,\vec{I}\,\vec{j}\,,\vec{I}\,\overleftarrow{j}\,)\cos^2(\omega t) + \left(\frac{\mathbf{B}^2}{2\mu_0}\right)_Z \overleftrightarrow{K}\sin^2(\omega t) \right] \right\} \tag{1.30}$$

and

$$m_{p_0}\vec{0} = \frac{V_m}{c^2}\left\{ \left[\frac{\varepsilon_0\mathbf{E}^2}{2}\right]_Y \vec{J}\,\overleftarrow{i} + \left[2\left(\frac{\varepsilon_0\mathbf{V}^2}{4}\right)_X (\,\vec{I}\,\vec{j}\,,\vec{I}\,\overleftarrow{j}\,)\cos^2(\omega t) + \left(\frac{\mathbf{B}^2}{2\mu_0}\right)_Z \overleftrightarrow{K}\sin^2(\omega t) \right] \right\} \tag{1.31}$$

in which (V_m= 1.497393267E-47 m^3) is the *maximum theoretical stationary isotropic volume* that the energy of the electron's intrinsic magnetic field reaches after having evacuated X-space during the mutual energy induction cycle that causes it to oscillate between constituting in alternance this magnetic field B and the neutrinic field v, which is an oscillation that replaces, in the structure of massive elementary particles ([31], [8] Chapter 11), the oscillation between the fields B and E characteristic of electromagnetic photons ([15], [8] Chapter 6) and of massive elementary particles carrier-photons ([31], [8] Chapter 11) ([32], [8] Chapter 14):

$$V_m = \frac{\alpha^5\lambda_C{}^3}{2\pi^2} = 1.497393267\text{E} - 47\,\text{m}^3 \quad \text{and} \quad \mathbf{v} = \frac{\pi(e')}{\varepsilon_0\alpha^3\lambda_C{}^2} \tag{1.32}$$

The neutrinic field v, that the trispatial geometry allows identifying for the first time, is introduced in Reference ([31], [8] Chapter 11) and is completely analyzed in Reference ([33], [8] Chapter 12), which also analyses the emission mechanics of neutrinos in the trispatial geometry. The *theoretical stationary isotropic volume* of energy of any elementary quantum was defined in Reference ([30], [8] Chapter 4).

During the decoupling process of an electromagnetic photon of 1.022 MeV or more, the energy in excess of the exact amount of 1.022 MeV that converts into the now invariant amount of energy constituting the separated masses of an electron and a positron, retains the LC structure of the incident double particle photon, but mechanically separates into equal parts between the two massive particles now separated as shown in Equations (1.27) and (1.28) and becomes their *carrier-photons*, propelling them in opposite directions in space at the velocity corresponding to the energy of their momentum, calculable with Equation (1.20), or with one of the following electromagnetic equations, developed in Reference ([42], [8] Chapter 5):

$$v = c \frac{\sqrt{\lambda_C (4\lambda + \lambda_C)}}{(2\lambda + \lambda_C)} \quad \text{or} \quad v = c \frac{\sqrt{4EK + K^2}}{2E + K} \tag{1.33}$$

A particular point of interest about the second equation (1.33) is that if the energy of the rest mass of the electron (E in the second equation) is reduced to zero (See Appendix A), only the energy of its carrier-photon remains in the equation, and its velocity can then only be the velocity of light, thus confirming the identity of its structure with that of de Broglie's double-particle photon ([15], [8] Chapter 6) ([42], [8] Chapter 5).

It is very easy to verify the validity of LC Equations (1.30) and (1.31) of the electron and positron, because all of their terms are very well known invariant physical constants. For example, by multiplying the maximum energy of the magnetic field in Equation (1.30) by the *theoretical stationary isotropic volume* of this amount of energy defined in Reference ([30], [8] Chapter 4), we effectively obtain half the energy of the invariant rest mass of the electron, which corresponds to its intrinsic magnetic field:

$$\frac{\mathbf{B}^2}{2\mu_0} V_m = \left(\frac{\mu_0 \pi e c}{\alpha^3 \lambda_C^2} \right)^2 \frac{\alpha^5 \lambda_C^3}{2\mu_0 \, 2\pi^2} = 4.09355206\ 8\text{E} - 14\ \text{j} \tag{1.34}$$

1.23. Construction of stable complex particles

It has been established long ago that all atoms are made of only three distinct types of stable subcomponents, electrons, protons and neutrons. All three are typically regrouped under the general term *"elementary particles"* in the community, that is, a *general* term that induces a certain amount of confusion because of these three, only the electron has been found to truly be an elementary particle, that is, that we have the experimental proof that it is not made of smaller subcomponents, but is demonstrably made exclusively of the electromagnetic energy that was the *substance* of the electromagnetic photon from which it emerged, as just put in perspective, and as analyzed in detail in Reference ([31], [8] Chapter 11).

The other two subcomponents of all atoms, the proton and the neutron, were found not to be charged and massive elementary particles of the same sense as the electron, but rather *systems of such elementary particles* in a state of stable stationary action electromagnetic equilibrium, just as the solar system is not a celestial body, but a system of celestial bodies stabilized in a state of stationary action equilibrium. Historically, the first suspicions that protons and neutrons were not really elementary particles were aroused by the difference in their behavior compared to that of electrons and positrons during the first non-destructive collision experiments between these particles in the first particle accelerators (**Figure 1.5**).

On their side, electrons and positrons always behave during mutual collision experiments as if they had at best a *point-like* presence in space, meaning that in their cases, unlike protons and neutrons, no seemingly unbreachable limit is detectable by collision at some distance from their centers, no matter how close two electrons or two positrons come to each other's centers during truly frontal collisions, which is a type of backward rebound seldom observed given that such frontal collisions between electrons or positrons are similar to bringing the highly sharpened tips of sewing needles into frontal collision (**Figure 1.6**).

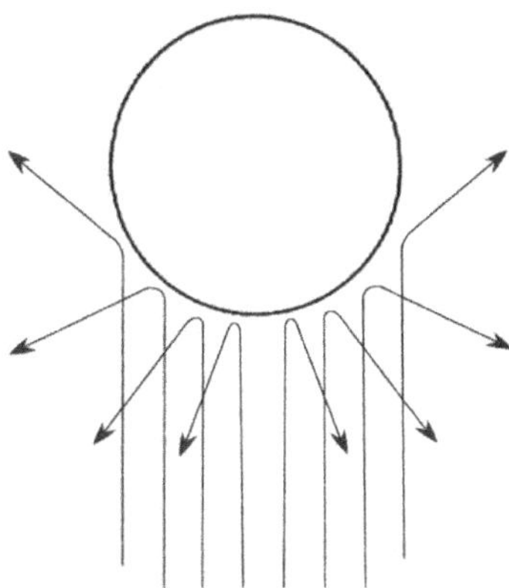

Figure 1.5: Perfectly elastic scattering between incident electrons and target proton.

It is this *quasi-punctual* or *point-like* behavior of truly elementary particles during mutual interactions or collisions experiments such as the electron, positron and electromagnetic photons that clearly differentiates them at the subatomic level from complex particles such as the proton and neutron.

What happened in the case of interactions between truly elementary charged particles was typically that incident electrons were deflected convergently as they crossed the position of positrons moving in the opposite direction, or when incident positrons crossed the path of electrons moving in opposite direction (**Figure 1.6-a**); or that incident electrons were deflected divergently after crossing the positions of other electrons moving in the opposite directions or when incident positrons crossed the position of other positrons moving in the opposite direction (**Figure 1.6-b**). Given the quasi-punctual behavior of the particles involved, only occasionally was one of the incident particles in an ideal situation to directly collide head-on in order to bounce back directly (**Figure 1.6-b**).

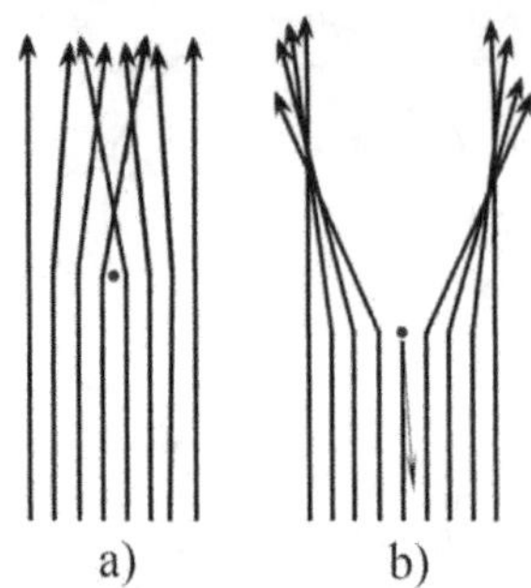

Figure 1.6. Non-destructive interaction between incident electrons and target positron a), and interaction and direct scattering between incident electrons and target electron b), demonstrating their point-like behavior.

While electron and positron beams launched so as to interact head-on with each other generated virtually no reverse rebounds (**Figure 1.6**), protons and neutrons caused the incident particles (electron or positron beams) to rebound in all directions (**Figure 1.5**), due to a state of permanent magnetic repulsion between the inner charged subcomponents of protons and the incoming electrons, as analyzed and described in Reference ([10], See also Section 2.21), which revealed that they occupy a measurable volume in space, contrary to electrons and positrons, that is, perfectly elastic rebound patterns identical to those that can be observed at our macroscopic level between two magnets repelling each other ([49], [8] Chapter 9).

The study of these rebound patterns in the 1940s and 1950s led to the conclusion that the radius of this volume was of the order of 1.2E-15 m for proton and neutron [57], a volume that seemed to reveal that they could be made of smaller particles whose interactions would determine this volume, just like the volume defined by planetary orbits determine the potential volume that the solar system can occupy in space, that is, theoretically at the time, truly elementary electromagnetic particles with *quasi-punctual* behavior of the same nature as the electron and the positron.

The first particle accelerator powerful enough to overcome the resistance of this proton volume to penetration by sufficiently energetic electrons or positrons, the Stanford Linear Accelerator (SLAC), came into service in 1966. From 1966 to 1968, a series of high energy non-destructive scattering experiments carried out by M. Breidenbach *et al.* [21] of electrons against protons effectively revealed the presence of three *quasi-punctual* behaving electrically charged subcomponents (**Figure 1.7**), whose deflection spread patterns of the incoming electrons' trajectories and subsequent analysis allowed associating an electric charge equal to 1/3 of that of an electron to one subcomponent and a charge equal to 2/3 that of the positron to the other two subcomponents (uud). Neutrons on the other hand revealed a structure made up of one 2/3 positive charge subcomponent and two 1/3 negative charge subcomponents (udd).

Moreover, incoming electrons backscattered in a highly inelastic manner and subsequent experiments also involving positrons revealed that the 2/3 positively charged subcomponents were only slightly more massive than electrons and that the 1/3 negatively charged subcomponent was only slightly more massive than the positively charged subcomponents ([32], [8] Chapter 14) [35].

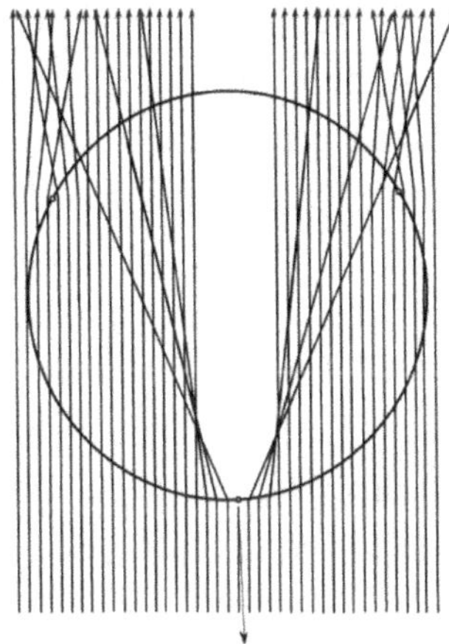

Figure 1.7: Detection of the proton inner structure via non-destructive scattering.

Given that these presumably invariant rest masses were eventually confirmed as being only slightly higher than those of electrons and positrons [51], combined with the fact that these sub-components of nucleons demonstrate exactly the same quasi-punctual behavior that characterizes electrons and positrons, and the also confirmed fact that electrons and positrons are the only massive and electrically charged elementary particles that can be generated from free electromagnetic energy in a well understood and exhaustively confirmed manner [23] [24], it seemed possible that these sub-components of nucleons could actually be positrons and electrons whose masses and charges would be altered in this way by the stresses imposed by those ultimate stationary action electromagnetic equilibrium states in which electrons and positrons could become captive of, if the latter truly are the only building material that nature has at its disposal to build nucleons.

This conclusion immediately explains why none of these nucleon sub-components could ever be ejected from a nucleon while still retaining its fractional charge, because if they really originally were electrons and positrons, then of course, they will naturally adiabatically recover their normal mass and charge characteristics as soon as they escape the electromagnetic stresses that they are subjected to while being part of the nucleon structures ([34], [8] Chapter 19).

The trispatial geometry indeed made it possible to calculate precise mean rest masses for these elementary positive and negative subcomponents of protons and neutrons corresponding to a sequence of stable axial resonance states that can be related to a sequence of integers, which locates these masses within the experimentally estimated possible mass ranges in both cases (See **Table 1.1**), that is, a sequence of three related masses that can be obtained from one of the possible equations that allows this calculation, such as the following equation established in Reference ([32], [8] Chapter 14), and analyzed in a more general perspective in Reference ([34], [8] Chapter 19), that is, a resonance sequence for the masses of stable elementary particles similar to the resonance sequence of the electronic orbitals of the hydrogen atom that Louis de Broglie was the first to notice at the beginning of the 20th century ([10], See also Chapter 2) [58]:

$$m_{i[d,u,e]} = \frac{k}{a_0}\left(\frac{3e}{n\alpha c}\right)^2 \quad (n=1, 2, 3) \tag{1.35}$$

where e is the unit charge, α is the fine structure constant, c is the speed of light, a_o is the Bohr radius, *i.e.* the mean axial distance between the fundamental electronic orbital of the hydrogen atom and the proton, and k is the Coulomb constant:

$$k = \frac{1}{4\pi\varepsilon_0} = 8.98755178\ 8E9 \tag{1.36}$$

Effectively, the masses obtained from Equation (1.35) fall right into the ranges experimentally established within which their true rest mass has to lie, that is, between 1 and 5 MeV/c^2 for the positive subcomponent, and between 3 and 10 MeV/c^2 for the negative subcomponent, according to experimentally obtained data on record [51]. These precise rest masses were established with respect to the distances separating the electromagnetically stressed electrons and positrons from the Y-z coplanar axis about which each stabilized triad is in *rotation/resonance* within electrostatic Y-space (**Figure 1.3**) as analyzed in Reference ([32], [8] Chapter 14).

The expression "*rotation / resonance*" is used here to clearly put in perspective that the same amount of energy is adiabatically induced by the Coulomb interaction in the rest masses of the electromagnetically stressed electrons and positrons, whether they are actually rotating on circular orbits about the Y-z coplanar axis an-d/or translation about the X-x normal axis, or simply are in a state of *stationary axial resonance* at these mean distances from these two Y-z and X-x *rotation / translation / resonance* mutually perpendicular axes.

Let us note, by the way, that at the time of the Breidenbach experiments [21], a mathematical theory developed separately by Murray Gell-Mann and George Zweig was considered confirmed by the Breidenbach experiments, which resulted in these electromagnetically stressed positrons and electrons captive of the nucleons' internal structures being named "*up quark*" and "*down quark*" respectively at a time when the conclusion had not yet been drawn that these nucleons' subcomponents could be simple positrons and electrons whose mass and charge characteristics could be altered in this manner by the intensity of the electromagnetic interactions at such short distances within these structures.

Since the Gell-Mann and Zweig theory also predicted the existence other virtual particles also named "*quarks*", but that never were detected by non-destructive collisions within nucleons, unlike the two that were named "*up*" and "*down*", the outcome was an enormous and persistent confusion in the community fuelled by multiple references to the Gell-Mann and Zweig theory, and the almost total absence of references to the experimental data gathered and analyzed by Breidenbach *et al.*, which left the impression during the following decades that even the sub-components actually detected by Breidenbach *et al.* were only theoretical and that their physical existence had never been confirmed.

Table 1.1: Sequence of masses in axial resonance state of elementary particles obtained using Equation (1.35).

	Rest mass	Energy	Charge	Ref.
Free moving electron or positron	9.10938188E-31 kg	0.511 MeV	±1= 1.602176462E-19 C	([31], [8] Chapter 11)
Electromagnetically stressed positron 1 in the neutron 2 in the proton	2.049610923E-30 kg	1.1497473 MeV	+2/3= 1.068117641E-19 C	([32], [8] Chapter 14)
Electromagnetically stressed electron 2 in the neutron 1 in the proton	8.198443693E-30 kg	4.59899 MeV	-1/3= 5.340588207E-20 C	([32], [8] Chapter 14)

The most edifying demonstration of this confusion is that in a major work on quantum field theory (QFT) published in 1993, that is, 25 years later, by a renowned physicist in the community, we find the following mention in Section 1.2 of his book [59], that shows that he had never heard about the Breidenbach *et al.* experiments that were carried 25 years before, otherwise it seems obvious that he would have taken them into account:

"Ironically, one problem of the quark model was that it was too successful. The theory was able to make qualitative (and often quantitative) predictions far beyond the range of its applicability. Yet the fractionally charged quarks themselves were never discovered in any scattering experiment."

However, in order to maintain continuity with the literature that was historically produced naming the electromagnetically constrained positrons and electrons as *up quarks* and *down quarks*, including the other articles of this series, we will keep the symbols "*u*" (for "*up*") and "*d*" (for "*down*"), that historically symbolized them when referring to the fractionally charged scatterable subcomponents of nucleons detected by Breidenbach, *i.e.* "*uud*" for the proton and "*udd*" for the neutron.

The trispatial LC equations for the electromagnetically stressed positrons (initially named "*up quarks*") and electromagnetically stressed electrons (initially named "*down quarks*") constituting the non-destructively scatterable nucleon structure are slightly different from Equations (1.30) and (1.31) that describe free moving electrons and positrons that are not being subjected to these electromagnetic stresses, because the transverse drift of the energy that defines the intensity of their fractional charges towards a more intense magnetic state, which is imposed on them by the very short gyroradius of their stationary action states ([60], [8] Chapter 8), does not allow an equal density of their electrical and magnetic states, unlike the default equal electric vs magnetic density state of the electromagnetic energy of electrons and positrons moving on straight line trajectories.

$$m_U = \frac{E_U}{c^2} = \frac{V_m}{c^2}\left\{S_U\left[\frac{\varepsilon_0 \mathbf{E}^2}{2}\right]_Y + (2-S_U)\left[2\left(\frac{\varepsilon_0 \mathbf{V}^2}{4}\right)_X \cos^2(\omega t) + \left(\frac{\mathbf{B}^2}{2\mu_0}\right)_Z \sin^2(\omega t)\right]\right\} \qquad (1.37)$$

$$m_D = \frac{E_D}{c^2} = \frac{V_m}{c^2}\left\{S_D\left[\frac{\varepsilon_0 \mathbf{E}^2}{2}\right]_Y + (2-S_D)\left[2\left(\frac{\varepsilon_0 \mathbf{V}^2}{4}\right)_X \cos^2(\omega t) + \left(\frac{\mathbf{B}^2}{2\mu_0}\right)_Z \sin^2(\omega t)\right]\right\} \qquad (1.38)$$

The expressions S_U and S_D are the *magnetic drift constants* of the energy of the stabilized rest masses of the electromagnetically stressed positrons and electrons, respectively equal to 2/3 and 1/3 and which are analyzed and explained in References ([32], [8] Chapter 14) and ([10], See also Chapter 2).

It is important to be aware that the sum of the stabilized rest masses of electromagnetically stressed electrons and positrons (**Table 1.1**) making up the scatterable structure of the proton (uud) constitutes only about 2% of its total measured mass, and that this sum for the neutron (udd) constitutes only about 2.4% of its total measured mass. The difference can only be due, of course, to the energy of their respective carrier-photons ([32], [8] Chapter 14), whose intensity depends directly on the inverse of the distance between charged elementary particles and the X-x translation axis of normal X-space (**Figure 1.3**) with respect to which each triad is in *translation/resonance*, an axis which is perpendicular to the coplanar *rotation/resonance* Y-z axis with respect to which the rest masses and fractional charges of the electromagnetically stressed electrons and positrons are determined ([32], [8] Chapter 14).

As in the case of the expression "*rotation/resonance*" previously mentioned in relation with the Y-space Y-z coplanar axis, the expression "*translation/resonance*" is used here to clearly put in perspective that the same amount of energy is adiabatically induced by the Coulomb interaction in each electromagnetically stressed electron and positron carrier-photon, whether they are in actual translation in circular orbit around the normal X-space X-x axis or simply are in a state of stationary axial resonance with respect to this mean distance from this translation/resonance axis, that is, a resonance motion oriented perpendicularly with respect to such a circular orbit.

1.24. The conceptual "translation/resonance" transposition

The same *translation/resonance* relationship also applies to the electron's rest orbital in the hydrogen atom for the same reason. In fact, it was Louis de Broglie who first understood in 1923 that the electron could only be in a state of axial resonance when stabilized at a mean distance from the proton in the hydrogen atom corresponding to the Bohr radius, even if it could also be theoretically perceived as being in translation on a closed orbit around the proton.

This conclusion of major importance was published in a note in which he proposed this first preliminary interpretation of the conditions that could explain the stability of the electron within atomic structures ([10], See also Chapter 2), since it was in harmony with the stability condition determined by Bohr and Sommerfeld for a trajectory traveled by a mass at constant velocity [58]. Here is a quote of his major conclusion:

> *"L'onde de fréquence v et de vitesse c/β doit être en résonance sur la longueur de la trajectoire. Ceci conduit à la condition:"*

Translation

> *"The wave of frequency v and velocity c/β must be in resonance over the length of the trajectory. This leads to the condition:"*

$$\frac{m_o \beta^2 c^2}{\sqrt{1-\beta^2}} T_r = nh \quad (n \text{ being an integer}) \tag{1.39}$$

It is this conclusion that gave Schrödinger the idea of representing the resonance volume visited by the electron in the rest orbital of the hydrogen atom by a wave function [18], as put in perspective in Chapter 2 that replicates the mechanical explanation of the hydrogen atom stability from the trispatial perspective initially published as Reference [10]. When de Broglie made his discovery, however, it was not yet clear that the very substance of the electron was truly electromagnetic in nature ([31], [8] Chapter 11), as well as that of its carrier-photon, that he intuitively identified as a *pilot wave* meant to propel the electron, but whose electromagnetic nature could not be identified at the time.

As mentioned earlier, it was not until the early 1930s that it was experimentally confirmed that the very substance of the invariant rest mass of the electron was nothing more than the *electromagnetic energy substance* of an electromagnetic photon of minimum energy 1.022 MeV decoupling into a pair of massive particles of equal masses, namely an electron and a positron [23]. Before this event, no one had had the opportunity to associate electromagnetic energy with the very substance of the mass of elementary particles, so none of the theories developed before this observation could take into account this new discovery in their elaboration, which of course includes Einstein's two theories of Special Relativity and General Relativity, as well as Quantum Mechanics in its traditional form.

De Broglie related the energy of the electron momentum at the Bohr orbit with Planck's constant and classical mechanics, but like the entire scientific community at that time, he did not relate it with the Coulomb interaction as represented with Equation (1.16) emerging from Maxwell's first equation and therefore he did not have at his disposal the conclusion that the half-quantum energy of the electron's momentum that would theoretically support the electron's motion longitudinally on its theoretical orbit around the proton is the same that also supports its axial resonance motion oriented perpendicularly to this orbit, as well as the associated half-quantum of its electromagnetic energy oriented transversely to this momentum energy, and that the unidirectional energy of its momentum can only be structurally oriented towards the proton.

In fact, the structural axial orientation of the momentum energy of the electron towards the proton does not exclude the possibility that the electron may move transversely on a closed orbit about the proton, in addition to oscillating simultaneously in axial resonance mode as de Broglie concluded, but at such a short distance between the electron and the proton and at such an intense level of induced energy, it can be expected that the axial resonance mode clearly dominates. See Sections 1.26 and 2.20.

It is a fact that the Planck constant associates the emission of electromagnetic energy strictly with the time factor. But this association of the induction of energy with the time factor is due to the fact that this constant was established by the analysis of the energy frequencies emitted during the de-excitation of electrons, that had previously been momentarily excited towards metastable orbitals further away from atomic nuclei, when they return to their rest orbitals of stationary action, which all are resonance states directly related to the frequency of the mean energy induced at the electron's rest orbital in the hydrogen atom, taken as fundamental, as analyzed and described in Reference ([34], [8] Chapter 19), and that the energy of Planck's quantum of action corresponds to the energy of a single cycle of this ultimate reference frequency, as subsequently determined by de Broglie:

$$h = m_0 v_B \lambda_B = 6.62606876\text{E}-34 \, \text{j}\cdot\text{s} \tag{1.40}$$

where m_0 is the rest mass of the electron, v_B is the conventional reference classical velocity of the Bohr orbit (2187691.253 m/s) and λ_B is the length of the Bohr orbit (3.32491846E-10 m), whose radius is the fundamental constant ($a_o=r_o=$5.291772083E-11 m), that is, the mean distance from the fundamental resonance orbital of the hydrogen atom to its nucleus, which defines the energy induced at this distance from the proton, or $E_B=$4.359743808E-18 j (27.21138346 eV) as easily calculated with the Coulomb Equation ([34], [8] Chapter 19). Its frequency is therefore $f_B=$6.579683921E15 Hz.

A simple calculation shows that at velocity v_B, the duration of a single cycle of this frequency corresponds exactly to the length of the Bohr orbit λ_B, which is why multiplying the length of this absolute reference orbit by the Planck constant makes it possible to obtain the energy induced at the Bohr orbit as precisely as with the Coulomb equation.

This is also why the energy corresponding to this reference frequency seems to correspond to the number of orbits that must be run in one second to supposedly *accumulate* all of the energy induced at the Bohr orbit, which has long created the perception that this induced energy *seems* to be distributed over all these cycles and that it takes one second for all the energy of the quantum to be accumulated:

$$E_B = h \cdot f_B = \frac{e^2}{4\pi \varepsilon_o r_B} = 4.359743808\text{E}-18 \, \text{j} \tag{1.41}$$

in which r_B is the Bohr radius, i. e. 5.291772083E-11 m (see Equation (1.7)).

Just as Marmet's Equation (M-7) can be generalized to use *the longitudinal electromagnetic wavelength* of any amount of electromagnetic energy, the same generalization was also made for the Coulomb equation in Reference ([30], [8]

Chapter 4), as analyzed and described in detail at Reference ([10], See also Chapter 2):

$$E = hv = \frac{e^2}{2\varepsilon_o \alpha \lambda} \tag{1.42}$$

where α is the fine structure constant (7.297352533E-3). The longitudinal wavelength of an amount of electromagnetic energy is also obtained using the following well-known equation, so the longitudinal electromagnetic wavelength of the energy E_B obtained with Equation (1.41) is:

$$\lambda = \frac{hc}{E_B} = 4.55633525 \, 2E - 8 \, \text{m} \tag{1.43}$$

which allows re-obtaining the same amount of energy with generalized Equation (1.42) already obtained with standard Equation (1.41):

$$E = hv_B = \frac{e^2}{2\varepsilon_o \alpha \lambda} = 4.359743808E - 18 \, \text{j} \tag{1.44}$$

It is in fact the relationship established with Equation (1.42) between the standard equation used to calculate electromagnetic photons energy and the generalized Coulomb equation that makes it possible to carry out the conceptual *translation/resonance* transposition required to be able to alternate between the analysis of the stable quantized energy states corresponding to all electronic and nucleonic stationary action orbitals in atoms, that relates Planck's constant with the number of theoretical cycles that the electron must theoretically run on the Bohr orbit; and that also allows the analysis of the infinitesimally progressive adiabatic induction of energy, which is a constantly active function of the inverse of the distance separating the charged elementary particles constituting all atoms, and which is induced *perpendicularly* by structure to any orbital motion, whether theoretical or effective.

This transposition in no way diminishes the usefulness of the Planck constant for calculations involving the study of the stable and metastable stationary action states of the various orbitals and the quantized emission of Bremsstrahlung photons, when de-exciting electrons move from a metastable orbital to a stable resonance orbital, whose emission mechanics we will analyze later, but it makes it possible to add to the body of mathematical tools the constants required to adequately deal with the infinitely progressive variations in the amount of energy adiabatically induced in electrons' carrier-photons by the Coulomb interaction during the axial resonance motion sequences into which they are captive when stabilized in the various stationary action orbitals in atoms, as analyzed in Reference ([10], See also Chapter 2), as well as when they are in free least action motion, *i.e.* in a process of moving towards these stabilized axial stationary action states, as analyzed at Reference ([43], [8] Chapter 2).

1.25. Electromagnetic energy adiabatic induction constants

1.25.1. The electromagnetic intensity constant

As analyzed and described in Reference ([30], [8] Chapter 4), since the speed of light is constant in vacuum, it can therefore be stated that the amount of energy of which an electromagnetic photon is made is inversely proportional to the distance it must travel in vacuum for one cycle of its wavelength to be transversely completed, which can be represented by $E=1/\lambda$. This means that by isolating product $E\cdot\lambda$ on the left side of this relation, the value obtained will be constant.

A quick analysis of Equation (1.44) reveals that this constant can alternatively be defined from the familiar set of electromagnetic constants that also defines the generalized Coulomb equation and the *longitudinal electromagnetic wavelength* of any amount of electromagnetic energy (λ):

$$H = E\lambda = \frac{e^2}{2\varepsilon_0\alpha} = 1.98644544 \text{ E} - 25 \text{ j} \cdot \text{m} \tag{1.45}$$

That is, the quantum of action in joules-meter (j·m), which is the counterpart dissociated from the time factor of the Planck quantum of action defined in joules-seconds (j·s), and that was named "*the electromagnetic intensity constant*" in Reference ([30], [8] Chapter 4). Dividing now constant H by the speed of light c, we observe that the Planck constant is obtained, which reveals that $H=hc$ directly links Planck's constant to electromagnetism, whereas it is historically considered to be strictly a measured constant not derived from electromagnetic equations:

$$h = \frac{H}{c} = 6.62606876 \text{ E} - 34 \text{ j} \cdot \text{s} \tag{1.46}$$

The unexpected result of this relation is that the Planck time based quantum of action can now be obtained from the same set of electromagnetic constants that defines constant H, by combining Equations (1.45) and (1.46), which makes available to the community this newly established definition of the Planck constant, established strictly from known fundamental constants and derived from experimentally confirmed equations, which is currently absent from both the "*CRC Handbook of Chemistry & Physics*" [51] and from the list of constants of the "*National Institute of Standards and Technology*" (NIST) [50]:

$$h = \frac{e^2}{2\varepsilon_0 \alpha c} = 6.62606876\,\text{E} - 34\,\text{j} \cdot \text{s} \tag{1.47}$$

1.25.2. The electrostatic energy induction constant

Metaphorically speaking, Planck's constant allows a *horizontal* (that is, *translational*) exploration of the stable orbital states of the hydrogen atom, so to speak, but the Coulomb Equation (1.41), which provides the same energy, was used to define *an electrostatic energy induction constant* that allows a *vertical* (that is, *axial*) exploration of the hydrogen atom and of its nucleus.

The required *electrostatic energy induction constant*, which was named K in Reference ([32], [8] Chapter 14) and could be considered as an *induction quantum*, was established in two different ways. The first method emerged from the analysis of the decoupling mechanics of a photon of energy 1.022 MeV into an electron-positron pair in the trispatial geometry, as established in Reference ([31], [8] Chapter 11), while the second method simply consists in multiplying Equation (1.41) by r_B squared:

$$K = E_B \cdot r_B^{\,2} = \frac{e^2 \cdot r_B}{4\pi\,\varepsilon_o} = 1.2208525596\text{E} - 38\,\text{j} \cdot \text{m}^2 \tag{1.48}$$

It is with this constant that it became possible to enter the hydrogen nucleus *vertically*, or *axially* so to speak, by varying distance r between two charged particles in equation $E = K/r^2$, and to thus establish the exact amounts of adiabatic energy induced in each of the internal components of the proton and neutron (see **Table 1.1**), thus allowing to finally establish coherent trispatial LC equations for the electromagnetically stressed electrons and positrons (see Equations (1.37) and (1.38) previously mentioned) and their carrier photons that determine the effective masses and the measurable volumes of protons and neutrons, as analyzed at Reference ([32], [8] Chapter 14).

1.26. Gravitation

In fact, such a *vertical* exploration, so to speak, of atomic and nuclear structures induces an acute awareness of the adiabatic nature of the energy induced in all of the charged particles making up their structures ([34], [8] Chapter 19) ([43], [8] Chapter 2), that is, an adiabatic energy that can only vary in an infinitesimally gradual manner with any variation in the distances separating them; an energy that moreover does not depend in any way on the velocity of particles, but that manifests its existence under the form of this velocity each time that local electromagnetic circumstances allow and that remains fully induced even if this velocity cannot be expressed due to the constraints imposed local electromagnetic equilibrium states.

As analyzed in References ([10], See also Chapter 2) and ([11], See also Chapter 3), when this velocity cannot be expressed, the momentum energy of each charged particle remains induced all the same and can then only exert a *pressure* in the vectorial direction imposed on it by the local electromagnetic equilibrium.

In atomic structures, this vectorial direction can only be towards the center of each atom due to the very nature of the Coulomb interaction. In accumulations of atoms making up larger masses, the tendency seems to be that this *pressure* tends to apply towards the centre of mass of these masses, which becomes obvious with masses such as that of the Earth, for example, on the surface of which all objects seem to be *attracted* to its centre of mass. But this supposed *attraction* can only be the *pressure* exerted by the total sum of the individual momentum energies of each charged particle constituting each object being applied against the surface of Earth, because their vectorial direction of application can only be by structure towards the Earth's centre of mass ([10], See also Chapter 2) ([11], See also Chapter 3).

In summary, the *weight* of an object as measured at the Earth's surface can only be a measure of this *pressure* exerted by the sum of the individual momentum energies vectorially oriented towards its centre of mass, belonging to the whole set of separate charged particles that constitute the measurable mass of this object. If this object is elevated above the ground and then left free to move, the velocity allowed by this sum of momentum energies can again be expressed until its motion becomes hindered again as the object meets again the surface of the Earth, at which point it will again exert a pressure equivalent to the amount of momentum energy induced by Coulomb interaction at this distance between each charged particle making up this object and each charged particle making up the Earth's mass ([43], [8] Chapter 2).

At the astronomical level, the celestial bodies of the solar system seem to be captive in stable stationary action resonance states at mean distances from the Sun similar to that which de Broglie assumed to apply to the electron in the hydrogen atom [58], *i.e.* a state of axial resonance limited by very precise minimum and maximum stable distances from the central star, that is, their perihelion and aphelion. These two boundary distances combined with the mean radius of the elliptical orbit of each celestial body constitute three stable references that allow clearly defining the volumes of space visited over time by each celestial body about the central star.

On the other hand, unlike the case of the hydrogen atom, as analyzed in Reference ([10], See also Chapter 2), for which the intensity of the momentum energy level induced in the electron at the mean distance from Bohr radius distance clearly favors a localized high frequency axial oscillation motion, rather than a translational motion along the theoretical Bohr ground orbit, the level of adiabatic energy induced in each charged particle of the Earth's mass at the average distance from the Earth's orbit being insufficient to generate such a high frequency axial oscillation, given the inertia of the macroscopic mass of which each charged particles is captive, rather favors the stabilization of celestial bodies in the observed states of stationary action orbital motion.

The volume of space visited over time by each celestial body about a central star can evolve into fairly complex shapes for celestial bodies that have satellites, which induces beat frequencies that cyclically modify the otherwise regular volumes visited by bodies that do not have a satellite. In fact, all bodies stabilized in such axial resonance systems mutually influence each other's trajectories and the shape of the resonance volumes they visit. It is this type of interaction, combined with the occultation process of the central star as these bodies pass between this star and our position in space that allowed the identification of the many planets orbiting nearby stars that have recently been discovered.

A similar electromagnetic dynamics defined by Quantum Mechanics (QM) is also applicable at the subatomic level to the elementary particles making up every atom of which all macroscopic masses are made, including our own bodies. In their cases, however, high-frequency axial stabilization, rather than orbital motion, is clearly favored due to the intensity of the adiabatic energy induced in each charged elementary particle at such short distances between the particles compared to their inertia.

An analysis initiated in References ([45], [8] Chapter 16) and ([61], [8] Chapter 15), and completed in Reference ([11], See also Chapter 3), of the sequence in decreasing order of intensities of the various stationary action states of electromagnetic equilibrium in which elementary particles can stabilize, shows that all possible cases of force application traditionally distributed among 4 fundamental forces: 1) *Strong interaction*, 2) *Weak interaction*, 3) *Electromagnetic force*, and finally 4) *Gravitational force*; can only be four quantized levels of Coulomb interaction intensity corresponding to the various energy levels of these stationary action equilibrium states.

Just like it seemed sensible to keep the terms "*up*" and "*down*" to designate positrons and electrons electromagnetically constrained within nucleon structures to maintain consistency with the bulk of previously published literature, it also seems sensible for the same reason to keep the easy to relate to concept of *attraction* to identify individual occurrences of Coulomb interaction between any pair of oppositely signed electrically charged particles. So, to facilitate the establishment of a mental image of the various orders of magnitude of electrostatic interaction application between any pair of such particles, the term "*attractor*" was defined in Reference ([45], [8] Chapter 16), embodying the idea that an *individual-inverse-square-of-distance-attractor* would be in action between each pair of these elementary particles in the universe. So, for simplicity's sake, any occurrence of the mentally easy to visualized concept of an electrostatic attraction between a pair of oppositely signed charged particles in the universe is referred to as an "*attractor*" in **Table 1.2**.

Table 1.2. Coulomb interaction quantized intensity ranges (See Reference ([45], [8] Chapter 16).

Table of electrostatic attractors		
Name	Range	Related Traditional force
Primary Attractor	Between electromagnetically stressed electrons and positrons inside a proton or neutron	Strong

Table of electrostatic attractors		
Name	Range	Related Traditional force
Secondary Attractor	Between electromagnetically stressed electrons and positrons belonging to different protons and neutrons in a nucleus	Weak
Tertiary Attractor	Between an orbiting electron and each electromagnetically stressed positron of an atomic nucleus and between each electron and each electromagnetically stressed positrons of other nuclei in all close by atoms	Electromagnetic
Temporary Local Attractor	Between half-photons inside a photon	Electromagnetic
Temporary Far Attractor	Between any half-photon and every other heterostatic particle in the universe	Electromagnetic
Quaternary Attractor	Between each charged particle in an atom and each heterostatic particle in relative free fall in the universe	Gravity

It now becomes possible to separate the Coulomb interaction gradient into four ranges of intensities, the boundaries of which correspond to the various ranges of stationary action resonance intensities that can be identified in nature (**Table 1.2**). As put into perspective in Reference ([45], [8] Chapter 16), the most intense level is determined by the resonance states characterizing the interacting electromagnetically constrained electrons and positrons forming the internal scatterable structure of nucleons, corresponding to the traditional *strong interaction*. The second level applies to the stabilization states of nucleons within atomic nuclei, corresponding to the traditional *weak interaction*. The third level applies to electronic resonance states within atoms and molecules, as well as between atoms and molecules in direct contact with each other in any accumulation of matter, corresponding to the traditional *electromagnetic force*. And finally, a fourth and final level of intensity applies to any atom, molecule and larger mass in a state of least action free fall, including those that are captive in stationary action orbits at the astronomical level, and corresponds to the traditional *gravitational force*.

These various levels of adiabatic carrier energy induction intensity by the Coulomb interaction, one of the major components of which is their transverse electromagnetic energy increment, corresponding to a variable increment of permanently induced adiabatic mass, provided for each charged particle in existence, can then be directly related with the 4 forces of the Standard Model as put into perspective in Reference ([45], [8] Chapter 16); four forces that ultimately turn out to be simple alternative representations of the various levels of intensity of application of a single *force*, namely the underlying adiabatic energy induction Coulomb interaction, as analyzed at Reference ([11], See also Chapter 3).

1.27. Nucleon expansion / compression as a function of the gravitational gradient intensity

The fact that the momentum half-quantum of adiabatic energy which is permanently induced by the Coulomb interaction in each electron is oriented axially towards the center of each atom taken separately, and that this energy can only be expressed as a pressure oriented towards the center of the atom when it cannot be expressed as a velocity, as analyzed and described in Reference ([10], See also Chapter 2), also has the consequence that when atoms accumulate to form larger masses, the vectorial resultant of all interactions between electrons and nuclei accumulated in close proximity will tend to orient the direction of application of these momentum half-quanta towards the centre of such masses, resulting in an addition of their individual pressures towards the centre of these masses.

When these accumulations of atoms become sufficient to form macroscopic masses, the resulting increase in pressure by addition as the depth increases in these bodies can only result in a forced contraction of the outer electronic orbitals of their atoms towards each their nuclei, as put into perspective in Reference ([45], [8] Chapter 16) and analyzed in depth in Reference ([43], [8] Chapter 2).

It is well verified that heat increases with depth in the Earth's mass [62]. However, it is also very well understood that heat in macroscopic masses is nothing more than an increase in the energy of the electrons of atoms, an increase which, when exceeding certain levels specific to each atom, forces the electrons of the outer layers of these atoms to jump to a metastable orbital further away from each nucleus involved. Since these levels are extremely unstable, these electrons return almost instantaneously to their stable stationary action orbital by then emitting a Bremsstrahlung photon that evacuates the energy (*i.e.* heat) accumulated as an electromagnetic photon, whose emission mechanics will be analyzed in the next section.

In the case of such heat increase with depth in planetary masses such as that of the Earth, it is well established that this increase is adiabatic in nature [62], and can only coincide with an adiabatic increase in energy by compression of the electronic orbitals towards their central atomic nuclei, because it is the resulting greater proximity between electrons and nuclei that causes the Coulomb interaction to induce this increased energy as a function of the inverse of the distance separating the electrons from the nuclei.

However, given that the atoms are in direct contact with each other in these masses and that this pressure is constant, this excess adiabatic energy cannot be evacuated by the emission of electromagnetic photons and simply increases with depth in the mass as the captive electrons of the outer electronic layers of the atoms approach the nuclei more and more as the depth increases, until an estimated temperature of about 5100 degrees Kelvin is reached at the centre of the Earth [62], as analyzed in Reference ([43], [8] Chapter 2).

Consequently, at the centre of proto-stellar masses in formation, for example, following a sufficient accumulation of interstellar hydrogen, this compression of the electron orbitals makes the hydrogen atoms electrons eventually reaches the distance to the proton that coincides with the induction of a carrier-energy in each electron reaching the critical decoupling threshold of 1.022 MeV for those at the very center of the proto-stellar mass, at which point decoupling into electron-positron pairs is forced by the immediate proximity of the high-frequency resonating charges of the proton, resulting in the formation of neutrons with enormous Bremsstrahlung energy emission that trigger and will subsequently maintain the nuclear fusion chain reaction in stars as analyzed in Reference ([45], [8] Chapter 16).

A side effect of the contraction of electronic orbitals towards nuclei with increasing depth in macroscopic masses such as planetary masses is that these atomic nuclei approach each other more and more as the depth increases in the mass, with these reduced distances intensifying the Coulomb interaction between the nuclei of these atoms.

The result is an increase in the outward *pull* involving the Coulomb interaction on all the charges of each nucleon in the various nuclei, which forces an increase in the *translation/resonance* distances of each triad relative to their central X-x axis of *translation/resonance* in X-space, decreasing the amount of adiabatic energy induced in their carrier-photons, thus decreasing the effective mass of all nucleons as depth increases in macroscopic masses, as analyzed in References ([32], [8] Chapter 14) ([45], [8] Chapter 16).

On the other hand, when small masses are taken away above the Earth's surface, the opposite effect can only occur by structure, because the energy of the electromagnetically constrained electrons' and positrons' carrier-photons of the nuclei of the atoms making up such small masses can only increase as a result of the increase in distances between them and all of the elementary charged particles making up the Earth's mass, which results in a contraction of the *translation/resonance* distances within each triad of such a small mass with respect to their normal X-x axis as a result of the weakening of the Coulomb interaction between the charges of these small masses and those of the Earth.

This contraction of the nucleonic orbitals within the nucleons of atomic nuclei making up such small masses moving away from the Earth, can only result in a proportional contraction of the electronic layers of these atoms, the measurable consequence of which is the increase in adiabatic energy induced at these shorter distances between the captive electrons and the nuclei, and therefore, in an increase in the electromagnetic frequency of the Bremsstrahlung photons emitted by electrons momentarily excited moving to a metastable orbital further away from the nucleus, as they de-excite almost instantaneously when returning to their stationary action orbitals.

Indeed, it can only be this mass increase of atomic nuclei with increasing altitude above the Earth's surface that really explains the increase in the frequency of Bremsstrahlung photons used in an atomic clock during the Hefele and Keating experiment [55] mentioned previously, supposedly demonstrating an alleged acceleration in the rate of *time* flow with altitude, then considered as a *proof* of the validity of SR ([45], [8] Chapter 16); which is a conclusion that was drawn before the adiabatic nature of the momentum energy and of the transverse magnetic field energy permanently induced in each charged elementary particle was put in perspective.

In reality, such atomic clocks, whose accuracy depends on the frequency of Bremsstrahlung photons emitted by electrons being de-energized, remain accurate as long as they are not moved from where they were calibrated. Any axial displacement in the gravitational gradient or change in its state of motion, such as when used in an orbiting satellite for example, requires recalibration that takes into account the local electromagnetic equilibrium.

Finally, the systematic *anomalies* observed about the trajectories of all space probes, extensively publicized in the case of the Pioneer 10 and 11 space probes about their escape trajectories from the solar system, that all behave systematically in deep space as if they were slightly more massive than measured on the ground before launch, also find a logical explanation in the previously analyzed fact that the rest masses of nucleons and macroscopic masses mandatorily vary by structure as a function of any axial displacement in the gravitational gradient.

There is then no doubt that the *anomalies* of the elliptical trajectories of Uranus, Neptune and Pluto, as well as those of comets Halley, Encke, Giacobini-Zinner, Borelli and others, that undergo systematic deviations of unknown origin as mentioned by R.W. Kühne [54], and in fact, all of the elliptical trajectories of the planets of the solar system, would benefit from being reconsidered with regard to this variability of their rest masses as a function of their axial oscillation in the Sun's gravitational gradient, and the variation of their transverse magnetic field as a function of their variable velocity on their elliptical trajectories.

1.28. The Bremsstrahlung photon emission mechanics

Now that the main conclusions that were drawn in the past about elementary particles, originating from already accumulated trustable experimental data, have been put in perspective in light of Maxwell's initial interpretation, the de Broglie hypothesis and Marmet's derivation within the broader framework of the trispatial geometry, let us now look at the Bremsstrahlung photon emission mechanics that this geometry allows establishing, that is, an emission mechanics that de Broglie and Schrödinger were looking forward to establish in the 1920s, but that elicited little interest in the community, for lack of a potential avenue of resolution to be explored at this time ([10], See also Chapter 2).

For this purpose, we will analyze the specific case of an electron in process of being captured by a proton to form a hydrogen atom, whose final stable least action equilibrium state, more precisely describable as state of *stationary* action, was analyzed in Reference ([10], See also Chapter 2). Before proceeding to the description of the actual emission mechanics, let us put some numerical figures in perspective with regard to the inertia of the various amounts of energy involved.

Immediately prior to its capture and stabilization at mean rest orbital distance from the proton (a_o=5.291772083E-11 m), the electron will have reached the relativistic velocity of 2187647.561 m/s, driven by the precise amount of ΔK momentum energy that its carrier-photon will have accumulated at this distance as it accelerated towards the proton ([43], [8] Chapter 2):

$$E_K = \Delta K = m_o c^2 (\gamma - 1) = 2.179784832\text{E-}18 \text{ j} \tag{1.49}$$

This velocity generates the *forward inertia* of the amount of momentum energy (13.6 eV) that will cause its own evacuation as an electromagnetic Bremsstrahlung photon as the forward motion of the electron is suddenly brought to a dead stop as a first step in the establishment of its stable axial stationary action orbital state. In addition to the forward inertia provided by this momentum energy, the total inertia of the incoming electron will also involve the inertia of the total amount of energy making up its carrier-photon transverse half-quantum and that of its invariant rest mass ($E=m_o c^2$=8.18710414E-14 j), both of which will not be evacuated during the stabilization process:

$$E_e = \Delta K + \Delta m_m c^2 + m_0 c^2 = 8.187540114\text{E} -14 \text{ j} \tag{1.50}$$

Equation (1.50) actually is the new trispatial energy-momentum equation that comes in replacement of the energy-momentum equation (2.41) traditionally associated with SR (see Section 3.5.1 and also Appendix A). On the other hand, the *stationary inertia* of the proton towards which the electron is accelerating depends on a much larger amount of energy:

$$E_p = m_p c^2 = 1.503277307\text{E-}10 \text{ j} \tag{1.51}$$

So the well known ratio of the inertias of both interacting components will of course be:

$$\frac{E_e}{E_p} = \frac{1}{1836054891} \tag{1.52}$$

We can observe that the forward inertia of the incoming electron is 4 orders of magnitude less than the stationary inertia of the proton, whose magnetic fields are its component that will stop the motion of the electron, by interacting in counter-pressure with respect to those of the incoming electron due to repulsive mutual parallel magnetic spin alignment imposed by structure, as clearly put in perspective in Reference ([10], See also Section 2.20). But the factual disproportion between the forward inertia of the electron momentum energy and the stationary inertia of the proton is immensely larger:

$$\frac{E_K}{E_p} = \frac{1}{6896448149} \tag{1.53}$$

This ratio reveals that whereas the forward inertia of the incoming electron will be countered by the stationary inertia close to 2000 times its own inertia, the forward inertia of the momentum energy of the incoming electron, that will be evacuated from the electron-proton system during the stoppage process, will be countered by a stationary inertia close to 69 million times its own forward inertia as the electron is coming in at a sizable fraction of the speed of light. This ratio puts in very clear perspective how instantaneously the forward motion of this momentum energy towards the proton will find itself countered during the stopping process.

However, contrary to the momentum energy of a moving object hitting a wall at our macroscopic level, for example, that we know experimentally will be communicated to the wall as the object hits it, we also know experimentally that the momentum energy of the incoming electron is not communicated to the proton, but will be ejected right out of the electron-proton system as a detectable and measurable outgoing electromagnetic photon of energy 2.179784832E-18 j, wavelength 9.113034513E-8 m and frequency 3.289710552E15 Hz, moving at the speed of light.

The issue of how the separation and ejection of this Bremsstrahlung photon mechanically proceeds has been pending ever since Louis de Broglie and Erwin Schrödinger began studying this process in the 1920's ([10], See also Chapter 2), but it was not really possible to resolve it before the expanded Maxwell compliant trispatial geometry previously described was elaborated and presented in 2000 at the event Congress-2000 [28].

This new space geometry now allows understanding that although the electron and its carrier-photon are suddenly stopped in their forward motion towards the proton while being abruptly captured at mean ground state orbital distance from the proton in a hydrogen atom, the forward motion of its ΔK momentum energy component calculated with Equation (1.49) is not stopped in its forward motion *within* the internal trispatial structure of the electron carrier-photon (**Figure 1.3-a and 1.3-b**), whose three separate spaces of its trispatial inner configuration act as communicating vessels ([15], [8] Chapter 6), that is, a forward inertia of the totality of the energy of electromagnetic photons which was confirmed by Einstein's photoelectric proof, i.e. in context $E=\Delta K+\Delta m_m c^2$.

The key to understanding why the motion of the ΔK momentum energy half-quantum of the electron carrier-photon is not stopped inside the carrier-photon as the latter is itself stopped in its forward motion, relates to step (c) of its trispatial electromagnetic cycle, as represented with **Figure 1.4**, which is the step, during the transverse oscillation cycle of the half-quantum $\Delta m_m c^2$, during which all of its transverse energy reaches its maximum volume within magnetostatic Z-space (**Figure 1.3**).

The manner in which the forward moving momentum energy ΔK of the electron being captured by the proton first crosses over to Z-space, as it own forward inertia forces it across the central point-like junction area interconnecting the three spaces through which the particle's energy freely transits within its own trispatial complex; and is then ejected backwards as a magnetic pulse during the electric phase of the carrier-photon's transverse oscillation cycle (**Figure 1.4-e**), as the two separated charges behave in Y-space, during the electron stopping process, as a fixed-length dipole antenna [63], can be summarized in a four steps sequence illustrated with **Figure 1.8**.

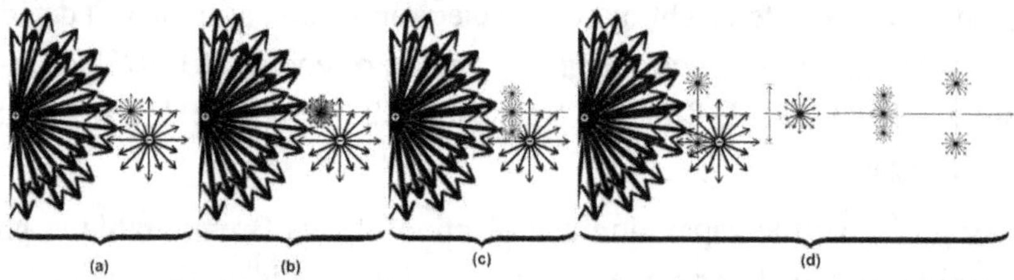

Figure 1.8. Representation of a Bremsstrahlung photon emission mechanics.

Figure 1.8-a represents the electron accompanied by its carrier-photon internally reaching step **1.4-c** (**Figure 1.4-c**) of its transverse oscillating cycle, as both of their magnetic fields begin colliding with the relatively huge magnetic field of the proton, as they repel each other by momentarily all being in parallel magnetic spin alignment, as analyzed in Reference ([10], See also Chapter 2).

Figure 1.8-b represents the second step of the ejection process, and illustrates the actual stopping sequence, as the complete complement of the ΔK=2.179784832E-18 J momentum energy has just been forced into Z-space by its own forward inertia, which actually momentarily doubles the amount of energy making up the magnetic field of the incoming carrier-photon, a doubling which is graphically represented by an increased visual density of the carrier-photon magnetic sphere:

$$2 \cdot \Delta \mathbf{B} = 2 \cdot \frac{\mu_0 \pi e c}{\alpha^3 \lambda^2} = 470103.4692 \, \text{T} \tag{1.54}$$

where λ=4.556335256E-8 m, which is the wavelength of the electron carrier-photon at the very beginning of the stopping process caused by the magnetic repulsion between its magnetic fields and that of the proton.

As it stands, this momentary doubling of the electron carrier-photon magnetic field as the electron begins to be captured in the hydrogen atom ground state should be detectable as a recordable magnetic intensity peak coinciding with the Bremsstrahlung photon emission, which would directly confirm the present photon emission mechanics.

Something else might already have drawn the reader's attention in **Figure 1.8-b**. Although the momentum energy initially belonging to X-space, which is represented by the left-pointing arrow leading to the carrier-photon magnetic sphere in **Figure 1.8-a**, was just mentioned as having been forced into Z-space by its own forward inertia to add up with the already residing magnetic energy as calculated with Equation (1.54), an identical arrow still is present in **Figure 1.8-b**. This requires an additional explanation, because this is no misrepresentation, because given that both the electron and the proton are electrically charged in opposition, the Coulomb interaction does not allow by structure that no momentum energy be induced in the electron carrier-photon at this distance from the proton, as put in perspective in Reference ([43], [8] Chapter 2).

Moreover, Reference ([52], [8] Chapter 10) clearly puts in perspective that a clear distinction must be made between an *uncompensated mechanically induced rotation or translation motion* and a *permanently compensated electrostatically or gravitationally induced rotation or translation*. Such *uncompensated* motion characterizes the state of a satellite launched into a metastable inertial orbit about the earth for example, or any object artificially rotated at our macroscopic level by means of a single initial impulse. The orbit of such an artificial satellite always degrades causing the satellite to crash, and the rotation of such an artificially rotated object always stops, unlike the natural *permanently compensated* orbit of the Earth for example, and its natural *permanently compensated* rotation. Considering the clear correlation previously established between translational or rotational motion and the states of stationary action resonance, the capture and stabilization of an electron in the stationary action resonance orbital of a hydrogen atom clearly belongs to the *permanently compensated* category, as put in perspective in Reference ([43], [8] Chapter 2).

Since the amount of ΔK momentum energy induced by the Coulomb interaction at this precise distance from the proton can in no way be different from 13.6 eV, it can be concluded that as the initial amount of forward moving ΔK momentum energy is evacuated from X-space into Z-space, a replacement 13.6 eV amount of ΔK momentum kinetic energy has to synchronously be adiabatically induced by the permanently acting Coulomb interaction in X-space, an energy whose vectorial direction of application will now be expressed as a *stationary pressure* exerted towards the proton, increasing, so to speak, the permanent counter-pressure established between the parallel-aligned magnetic fields involved ([10], See also Section 2.20). This means that momentarily, the carrier-photon will involve 40.8 eV of energy, including now the momentary double intensity magnetic field, until the 13.6 eV temporarily transferred to Z-space is subsequently evacuated as a separate out-going electromagnetic photon in the following manner.

Figure 1.8-c represents the setting up of the metaphorical dipole antenna that will emit the excess 13.6 eV energy as an electromagnetic photon. As the carrier-photon magnetic field reached maximum *presence* in Z-space as represented in **Figure 1.8-b**, the related dipole electric field was down to zero *presence* in the carrier-photon Y-space, which corresponds to the two rods of a fixed length dipole antenna being neutral when no alternating current is provided to the antenna [63].

As the magnetic energy represented in **Figure 1.8-c** starts naturally moving back into the electrostatic carrier-photon Y-space, it builds up in this space as two opposite charges moving in opposite directions on the Y-y/Y-z plane ([15], [8] Chapter 6) ([34], [8] Chapter 19), causing the two opposite charges to eventually peak at their maximum allowed value, which, at this precise moment, cannot exceed the maximum transverse E-field energy authorized mean value of 2.179784832E-18 J (13.6 eV) at this distance between the positively charged proton and the negatively charged electron, which combined with the newly induced equal authorized momentum energy value which is now *stationarily pressuring* the electron against the magnetic field of the proton, and is adiabatically maintained by the Coulomb interaction at this mean distance.

It is this maximum E-field energy limit enforced by the Coulomb interaction that causes the sudden maximizing of the distance between both charges in Y-space causing it to momentarily act similarly the two fixed length rods of a dipole antenna, that allow the extra energy that was forced into Z-space, initially coming from X-space, to now moves on into Y-space and overload the now momentarily fixed maximized length of the Y-space dipole, causing it to emit the excess 13.6 eV energy as a magnetic pulse in magnetostatic Z-space, in the same manner as electromagnetic energy pulses are emitted from a very normal dipole antenna at our macroscopic level, which is represented with **Figure 1.8-d**.

The question comes up here as to why does the electron not simply fly away from the proton since it is universally known to do so when precisely this amount of ΔK=2.179784832E-18 j energy that it now already possesses is provided to it from an incoming electromagnetic photon, which is the case that will be addressed in the next and last Section of this chapter. The answer is really simple in this specific case, and is provided by simply becoming aware that the whole practically instantaneous sequence represented by **Figure 1.8** occurs while the *forward inertia* of the total amount of energy making up the electron invariant rest mass and its carrier-photon is applying its maximum pressure against the magnetic field of the proton, momentarily defeating any possibility for the electron to be ejected at this precise moment, and also defeating any possibility for the distance between the electron and the proton to vary during this so brief stopping sequence process.

Right after having been chased into Z-space by the Y-space electric dipole, the first thing that will happen to the expelled energy will be the transfer from Z-space to X-space of half its energy within a new trispatial set of communicating vessels, to build the momentum energy half-quantum that will then start propelling the fledgling photon at the speed of light away from the proton, as the first step of the re-establishment of its natural trispatial electromagnetic equilibrium. Once both energy half-quanta have reached their default equal longitudinal and transverse energy levels as could be determined according to de Broglie's hypothesis and from Marmet's derivation, the energy of its transverse magnetic *B*-field will naturally start transversely oscillating by crossing over to Y-space to induce the corresponding *E*-field, thus initiating the stable transverse electromagnetic oscillation of the new Bremsstrahlung photon, now moving freely at the speed of light, as represented with **Figure 1.8-d** ([15], [8] Chapter 6).

Note that although the complete process took a noticeable amount of time to describe, the actual sequence of events causing the electron to come to a momentary dead stop as it is being captured by a proton, has to be practically instantaneous, due to the velocity of the incoming electron, combined with the fact that the whole sequence definitely has to be completed during the fleeting half-cycle of the carrier-photon transverse electromagnetic oscillation, beginning with its parallel magnetic spin alignment (**Figure 1.4-c**) with respect to the spin orientation of the magnetic field of the proton and ending with the maximum *E*-field charges separation (**Figure 1.4-e**) as represented at the beginning of **Figure 1.8-d**; the whole sequence occurring, as previously mentioned, while the inertia of the total amount of energy making up the electron invariant rest mass and the momentarily invariant mass of its carrier-photon are applying maximum pressure against the magnetic field of the proton ([10], See also Section 2.20).

1.29. The electromagnetic photon absorption mechanics

As soon as the Bremsstrahlung photon has been emitted, the *forward inertia* of the electron invariant mass/electromagnetic-fields and of its carrier-photon variable mass/electromagnetic-fields half-quantum, due to their incoming velocity, will be replaced by their default *stationary inertia*, to which must be added the *adiabatically variable forward pressure* provided by the newly induced *ΔK* carrier-photon momentum energy half-quantum, which is permanently oriented towards the proton, that jointly interacts in counter-pressure with respect to the *oscillating*, but nevertheless *stationary inertia* of the much larger mass/electromagnetic-fields of the proton, which interaction establishes and maintains the electron on its axial least action resonance trajectory within the stationary action volume of space that Schrödinger meant to describe with the wave equation [18], as described in Reference ([10], See also Section 2.20).

Now that only the permanent *forward pressure* of the recently adiabatically induced ΔK momentum energy is preventing the electron from escaping, and that the *momentary pressure* that was initially exerted towards the proton due to the *forward inertia* of the electromagnetic fields of the electron and carrier-photon, which initially prevented the electron carrier-photon transverse E field energy from exceeding its incoming initial value of 2.179784832E-18 j, is no more in action, after having caused the Bremsstrahlung photon to be emitted, as just analyzed; any energy coming from outside the electron-proton system can now be captured by the Y-space electric dipole of the carrier-photon, presumably still acting as a dipole antenna, but whose length can now vary, and will be equally distributed between both carrier-photon half-quanta, to the extent that the electron's magnetic gyroradius in the hydrogen atom will allow ([60], [8] Chapter 8).

The resulting increase in the axial resonance volume that the electron will visit as a result, will cause the electron to eventually jump to an authorized metastable orbital further from the proton before returning almost immediately to the rest orbital, emitting in the process a Bremsstrahlung photon that will evacuate the corresponding excess energy, or to escape completely from the proton if the energy supplied from outside the electron-proton system reaches the escape level of ΔK=2.179784832E-18 j, either by progressive accumulation or by collision with an incident photon of energy 2.179784832E-18 j.

All possible cases of energy emission and absorption must of course be explained and documented in context of the trispatial geometry, but since this document is intended only to put in perspective the underlying electromagnetic context that allows a general description of the mechanics of electromagnetic photon emission and absorption by electrons in the trispatial geometry, as a complement to the establishment of the electron stabilization mechanics in the hydrogen atom previously described in Reference ([10], See also Chapter 2), their development is beyond the scope of the paper reproduced in this chapter.

1.30. Conclusion

This analysis highlights the point that it is no more difficult to conceive that electromagnetic energy can consist of localized photons at the subatomic level than it is to conceive that water consists of localized molecules at the submicroscopic level, even if at our macroscopic level we treat electromagnetic energy as if it was made of continuous wave impulses and water as if it was a fluid without internal structure.

The main conclusion of this paper, however, is that when Maxwell's initial interpretation is correlated with the de Broglie hypothesis about the double-particle photon and Marmet derivation in context of the trispatial geometry, electromagnetism can finally be completely harmonized with Quantum Mechanics, as analyzed in Reference ([10], See also Chapter 2); a harmonization that now allows a first mechanical explanation to the processes of electromagnetic photon emission and absorption by electrons, as previously described.

It must also be clearly put in perspective that Maxwell's initial interpretation is a conclusion firmly grounded on the study and analysis of experimental data collected earlier during easily reproducible experiments that were performed by many experimentalists, as well as on the conclusions and equations that they drew from this data. The electromagnetic equations generally referred to as "*Maxwell equations*" are in fact a set of equations that Maxwell established as complementary, but that were developed mainly by Coulomb, Gauss, Ampère and Faraday (See Appendix B). Lorentz, Biot, Savart and a few others then completed the current set of mutually complementary electromagnetic equations from the analysis of more data obtained from other experiments equally easy to reproduce.

Intrigued at not finding any evidence of an experiment confirming the point-like magnetic behavior of spherical magnetic fields whose two poles coincide geometrically, which must be the *de facto* magnetic structure of electrons, given their systematic point-like behavior during all scattering experiments, this author designed and carried out in 1998 an easily reproducible experiment with magnets magnetized accordingly, whose data and subsequent analysis were published in 2013, for the experiment to become available in the education community ([49], [8] Chapter 9). One year later, S. Kotler *et al.* published an article describing an experiment performed with actual electrons that directly confirmed the prediction of the 1998 experiment [64].

The education community now has at its disposal a complete set of demonstration experiments easily reproducible during hands-on laboratory teaching sessions, ranging from the first Coulomb electric experiment to the 1998 magnetic experiment to help teaching and confirming every aspect of electromagnetic energy behavior.

2. The Hydrogen Atom Fundamental Resonance States

2.1 Introduction

In the 1920's, Louis de Broglie's observation that the integer sequence that could be related to the interference patterns produced by the various electromagnetic energy quanta emitted by hydrogen atoms was identical to those of very well known classical resonance processes, made him conclude that electrons were captive in resonance states within atoms. This led Schrödinger to propose a wave function to represent these resonance states that still have not been reconciled with the electromagnetic properties of electrons. This article is meant to identify and discuss the electromagnetic harmonic oscillation properties that the electron must possess as a resonator in order to explain the resonance volume described by the wave function, as well as the electromagnetic interactions between the elementary charged particles making up atomic structures that could explain electronic and nucleonic orbitals stability. An unexpected benefit of the expanded space geometry required to establish these properties and interactions is that the fundamental symmetry requirement is respected by structure for all aspects of the distribution of energy within electromagnetic quanta.

This paper does not propose an alternate approach to quantum mechanics, but rather an addition to the already established descriptions of orbital resonance states provided by Schrödinger's wave function, Heisenberg's statistical distribution and Feynman's path integral, involving a clear description of the electromagnetic resonators responsible for the establishment of the related resonance volumes, meant to lay the groundwork for the eventual establishment of more elaborate wave functions that will for the first time completely account for the electromagnetic nature of these resonators.

The detailed mathematical proof of complete conformity with electromagnetism and with every aspect of all experimental data on record is provided in a series of articles previously published that are given in reference as required. This new approach is in complete agreement with the methods of QED and QFT and complements them by clarifying the function of the magnetic aspect of the energy of which electromagnetic elementary particles and their carrying energy is made, in a manner that allows describing their permanently localizable self-sustaining internal electromagnetic structure.

The key concept that triggered this particular research is an aspect of the wave function that seems to have escaped general attention almost from the moment that Quantum Mechanics (QM) was established to represent the hydrogen ground state from the correlation of Heisenberg's statistical representation and Schrödinger's wave function. This concerns the very reason why Schrödinger came up with the idea of using a wave function to describe the already well known stable ground state of the electron in the hydrogen atom. Strangely, it seems that the seminal paper which is at the origin of this major discovery never was translated to English to be put at the disposal of the international community [58].

This paper, written by Louis de Broglie, relates the interference patterns produced by the various electromagnetic energy frequencies emitted by hydrogen atoms to possible resonance states of the electron on what was then perceived as the various orbits that it could occupy in the hydrogen atom.

Here is de Broglie's description in his own words of the observation that led him in 1923 to this major conclusion, followed by its apparently first time ever translation to English:

> *"L'apparition, dans les lois du mouvement quantifié des électrons dans les atomes, de nombres entiers, me semblait indiquer l'existence pour ces mouvements d'interférences analogues à celles que l'on rencontre dans toutes les branches de la théorie des ondes et où interviennent tout naturellement des nombres entiers."* ([4], p.461).

> *"The occurrence of integers, in the laws of quantified motion of electrons in atoms seemed to me indicative of the existence for these motions of interferences analogous to those met in all branches of waves theory, where integers naturally occur".*

Shortly afterwards, he published a note in the *Comptes rendus de l'Académie des Sciences* in which he was proposing the first preliminary interpretation of the conditions that might explain the stability of the electron within atomic structures [58].

The critical conclusion of this note is the following. Original in French:

"L'onde de fréquence v et de vitesse c/β doit être en résonance sur la longueur de la trajectoire. Ceci conduit à la condition:"

"The wave of frequency v and velocity c/β must be in resonance on the whole length of the trajectory. This leads to condition:"

$$\frac{m_o \beta^2 c^2}{\sqrt{1-\beta^2}} T_r = nh \qquad n \text{ being an integer} \tag{2.1}$$

which is the stability condition determined by Bohr and Sommerfeld for a trajectory being run at constant velocity [58].

The following year, de Broglie published two more notes, in one of which he mentioned that from this viewpoint, Bohr's famous *frequency condition law* could be interpreted as involving some sort of *beat* or *pulsation* ("*un battement*" in the original French text), that is, a resonance state associating the frequency of the emitted wave to the initial electron stationary state and to its final stationary state ([4], p. 462), [65] and [66]. Two years later Schrödinger introduced the wave function to account for this measurable condition.

The obvious starting point of his exploration was a complex-valued simple harmonic oscillation formula that then evolved into more elaborate spherical formulations to describe the ground state orbital of the hydrogen atom [18].

To this author's knowledge, no subsequent mention of the fact that Schrödinger's wave function is meant to describe a stable resonance state into which the localized electron remains captive can be found in the historical formal literature, except in a book published in 1953 to which the actual discoverers of Wave Mechanics and Quantum Mechanics collaborated [4].

Moreover, although Einstein contributed the text of the introduction of this book in German, and that Schrödinger contributed in English the Chapter that he provided, both contributions being translated to French on the facing pages, the remainder of the book was in French only. It also appears that this particularly important book, to which Einstein, Schrödinger, Pauli, Rosenfeld, Heisenberg, Yukawa, Davisson and de Broglie, to name only the most famous, and many more, jointly collaborated to provide a general overview of the state of quantum physics as of 1952, highlighting in historical context the contribution of Louis de Broglie, was apparently never translated to English nor to any other language to be made available to the international scientific community.

From what can be learned from this book, soon after the wave function was introduced by Schrödinger, whose validity was confirmed within a few years as being incontrovertibly related to resonance states, according to interference patterns generated during experiments carried out by Davisson and Germer, and also by G.P. Thompson ([4], p.19), the adoption by the majority of researchers of Heisenberg's statistical representation, that replaces the volume of isotropic energy density defined by Schrödinger's wave function by a distribution of the energy density of the electron according to a statistical arrangement reflecting an *amplitude probability* perceived as being a more precise representation than the initial wave function, giving a greater density of energy presence in the vicinity of the Bohr radius for example, caused the fact that the wave function was initially meant to represent a resonance state to be obscured and neglected practically from the onset.

The probabilistic interpretation also favored the idea of sudden jumps from one energy level to another, providing no mechanical explanation to these jumps, contrary to the wave equations that had the potential to allow descriptions of such changes as being mechanically progressive and mathematically describable processes, as re-emphasized by Schrödinger in 1953:

> *"To produce a coherent train of light waves of 100 cm length and more, as is observed in fine spectral lines, takes a time comparable with the average interval between transitions. The transition must be coupled with the production of the wave train... For the emitting system is busy all the time in producing the trains of light waves, it has no time left to tarry in the cherished stationary states", except perhaps in the ground state."* ([4], p.18).

Even Einstein, who, like de Broglie and Schrödinger, was convinced that the electron remains constantly localized as it moves and always follows a precise trajectory, was unconvinced by the discovery by de Broglie's of the relation between discrete quantum states and resonance states, presumably because he did not associate the concept of *mass* to electromagnetism in the same manner as de Broglie and Schrödinger.

Here is Einstein's comment in this regard that appears at the beginning of the introduction of this book. Original text in German:

> *"Ich will dem zusammen mit Frau B. Kaufman verfassten Beitrag zu diesem Bande einige Worte vorausschicken in der einzigen Sprache, in der ich mich mit einige Leichtigkeit ausdrücken kann.*

Es sind Worte der Entschuldigung. Sie sollen zeigen, warum ich, trotzdem ich De Broglie visionäre Entdeckung des inneren Zusammenhanges zwischen diskreten Quantenzuständen und Resonanzzuständen in relativ jungen Jahren bewundernd miterlebt habe, doch unablässig nach einem Wege gesucht habe, das Quantenrätsel auf anderem Wege zu lösen oder doch wenigstens eine Lösung vorbereiten zu helfen." ([4], p. 4).

Translation:

"I will begin my contribution prepared for this book in collaboration with Mrs. Kaufman with a few words in the only language in which I can express myself with any ease. They are words to express regret. They are meant to show why – although I observed admiringly in my years of relative youth the genial discovery by Louis de Broglie of the intimate relationship between the discrete quantum states and resonance states – I nevertheless ceaselessly searched for some manner to resolve the enigma of quanta by some other means, or at least help in preparing such a solution."

It turns out that Schrödinger and de Broglie were initially analyzing these observed resonance states in view of establishing a progressive mechanical explanation to the transitions between the stationary states, that would explain the generation of the Bremsstrahlung photons responsible for the fine spectral lines detected in relation with these transitions (See Section 1.28), but that the immediate popularity of Heisenberg's statistical method in the community caused all research in this direction to be stalled almost from the start.

Schrödinger clearly expressed his frustration for the neglect of any research in this direction in the chapter that he contributed:

"For it must have given to de Broglie the same shock and disappointment as it gave to me, when we learnt that a sort of transcendental, almost psychical interpretation of the wave phenomenon had been put forward, which was very soon hailed by the majority of leading theorists as the only one reconcilable with experiment, and which has now become the orthodox creed, accepted by almost everybody, with a few notable exceptions." ([4], p. 16).

Schrödinger and de Broglie were obviously convinced that the frequency of an emitted quantum could be produced only by means of a progressive mechanical process dependent on the resonance characteristics of an electron initial stationary state, and whose emission mechanically determined in a clearly describable manner the altered resonance characteristics of the final stationary states and that solving this problem would be useful not only in spectroscopy, but also in chemistry. See Sections 1.28 and 1.29 for the mechanics of electromagnetic photons emission and absorption as definable from the trispatial perspective.

It seems that Schrödinger's frustration was well justified, considering that it took 55 years after he so openly aired this protest in this book and also in a paper titled "*Are there quantum jumps*" published the same year in the "*British Journal for the Philosophy of Science*" [67], that is, 80 years after he introduced the wave function, for the first hints of renewed interest in resonance states in relation with the wave function to reappear in the community. This recent analysis can be found in a paper by V. A. Golovko [68] published in 2008.

The outcome of the adoption by the majority of theorists of the statistical method as representing fundamental reality then led to the establishment of the quantum field theory (QFT) grounded on an axiomatic fundamental concept of spontaneous quantum energy fluctuations on either side of a zero point energy level that would exist everywhere in space, that establishes virtual photons (bosons) as being the force carriers that would explain the energy levels and motion of real elementary electromagnetic particles in space. See Sections 3.1, 3.11 and 3.27.

These hypothetical spontaneous stochastic fluctuations of the underlying quantum field are also considered as explaining an apparently erratic transverse quivering motion observed in the behavior of moving electrons in certain circumstances that Schrödinger named "*zitterbewegung*" ("*trembling motion*"), which we will analyze further on [69]. See Section 2.18.

It is quite obvious that QFT is correctly grounded on Maxwell's electromagnetic wave theory and equations, but it nevertheless obscures the fact that in electromagnetism, an electron, for example, which is electrically charged, can be made to move in a straight line when immersed in equal density ambient E and B fields; that if these intensities are made to simultaneously gradually vary, even if the variation is infinitesimally progressive, its velocity will vary just as gradually, and if their relative densities are made to gradually differ from each other, this will cause the electron to just as gradually curve its trajectory, which are processes all aspects of which can be calculated and controlled with the Lorentz Equation $(F = q(E + v \times B))$ *(See Section B.3).*

This behavior of electrons completely validates the possibility that if the QFT concept of virtual bosons being the force carrier was replaced by the infinitesimally progressive Coulomb interaction stemming from Maxwell's first equation, that is, Gauss's equation for the electric field, this opens up the possibility that electromagnetic Bremsstrahlung photons could be defined as self-sustaining their own motion in a localized manner without the need for any underlying aether, from the simple interaction of their own internal mutually inducing E and B fields in conformity with Maxwell's grounding hypothesis, and that they could be defined as default self-guiding in straight line from the default equal densities of there own internal E and B fields ([15], [8] Chapter 6).

2.2. The E and B fields of the moving electron

It can also be observed that the resonance states of the electron are not the only aspects of electrons that seem to have been the object of little research over the course of the past century.

Despite the known facts that the electron possesses and electric charge, that it can be guided by progressively varying ambient electric E and magnetic B fields and that the *wave* aspect of its established *wave-particle* nature confirm that it is an electromagnetic particle, it seems that the intrinsic E and B fields of the electron itself, that is the E and B fields that must be associated to its very charge and mass, have apparently not yet been investigated in the community.

Indeed, the only relations between the electron and E and B fields that can apparently be found in the literature of the past hundred years specifically refer to the motion of electrons in ambient electric or magnetic fields, without any mention of any kind of interaction between these external fields and those that have by structure to be related to the electric charge and rest mass of the electron.

The first breakthrough in this direction is fairly recent. In 2003, Paul Marmet succeeded in directly relating the increasing magnetic field of an accelerating electron to its increase in relativistic mass from quantizing the electron charge in the Biot-Savart equation [29].

After he established the electron charge as remaining invariant at its unit value (1.602176462E-19 C) in the Biot-Savart equation, his Equation (M-17) now provides us with an electromagnetic equation that allows directly calculating *the mass increment corresponding to the magnetic field increment* of the accelerating electron:

$$\Delta m_m = \frac{\mu_0 \left(e^-\right)^2}{8\pi r_e} \frac{v^2}{c^2} \tag{2.2}$$

This equation thus directly associates the concept of *classical mass* to the real electromagnetic energy that must by definition be associated with this *magnetic field* increment of the electron in motion, which involves by similarity that the intrinsic magnetic field of the electron must also be associated to the real electromagnetic energy making up its invariant rest mass, as we will shortly see.

He also observed that since the varying inertial mass of the moving electron is given by

$$m = \gamma m_e \tag{2.3}$$

and that the Lorentz γ factor can be expanded as the following series:

$$\gamma = 1 + \left\{ \frac{1v^2}{2c^2} + \frac{3v^4}{8c^4} + \frac{5v^6}{16c^6} + \frac{35v^8}{128c^8} + \cdots \right\} \tag{2.4}$$

and that the term $(v/c)^4$ and other higher order terms are negligible with respect to the $(v/c)^2$ term, they can be ignored for low relativistic velocities, which allows establishing the following equality from Equation (2.4):

$$\gamma\text{-}1 = \frac{1}{2} \frac{v^2}{c^2} \tag{2.5}$$

Knowing that the momentum related relativistic kinetic energy of a moving electron is obtained from the following standard equation, which makes use of the right term of Equation (2.5):

$$\Delta K = m_0 c^2 \left(\gamma - 1\right) \tag{2.6}$$

we can similarly calculate its relativistic mass increment from combining Equation (2.3) and Equation (2.5):

$$\Delta m = m\text{-}m_e = m_e(\gamma - 1) = \frac{m_e}{2}\frac{v^2}{c^2} \tag{2.7}$$

Comparing now Equation (2.2) with Equation (2.7), we observe that we now have two different equations representing the same mass increment of the moving electron; that is, Equation (2.2) providing this increment as the mass of the magnetic field increment, while Equation (2.7) provides this same increment as a *classical mass* increment. We can thus equate Equations (2.2) and (2.7) in the following manner:

$$\Delta m_m = \Delta m = \frac{\mu_0\left(e^-\right)^2}{8\pi r_e}\frac{v^2}{c^2} = \frac{m_e}{2}\frac{v^2}{c^2} \tag{2.8}$$

and finally, when the velocity becomes infinitesimal, both velocities ratios can be ignored to finally reveal the surprising fact that the mass of the intrinsic magnetic energy of the electron makes up exactly half of its invariant rest mass, which is the critically important conclusion reached by Marmet:

$$m_m = \frac{\mu_0\, e^2}{8\pi\, r_e} = \frac{m_e}{2} \tag{2.9}$$

2.3. The electron carrying energy

Let us consider for a moment the meaning of Δm_m from Equation (2.2) and of ΔK from Equation (2.6). To be able to really see what is involved, let us use the well known concrete case of the electron relativistic velocity of 2187647.561 m/s on the theoretical classical Bohr ground state orbit. Using this velocity to resolve Equation (2.2), we obtain the following mass increment:

$$\Delta m_m = \frac{\mu_0\left(e^-\right)^2}{8\pi r_e}\frac{v^2}{c^2} = 2.425337715\mathrm{E}-35\,\mathrm{kg} \tag{2.10}$$

which is the magnetic field mass increment to be added to the rest mass of the electron to obtain the total effective electron mass that experimentalists have to deal with when transversely deflecting electrons moving freely at this corresponding relativistic velocity of 2187647.561 m/s.

Multiplying now this value by c^2, we obtained the energy in joules constituting this amount of mass (2.179784832E-18 j), and further dividing this value in joules by the unit charge of the electron (1.602176462E-19 C), we obtain its conversion to electronvolts (13.6 eV).

Let us now calculate with Equation (2.6) the momentum kinetic energy related to this same velocity of the electron:

$$\Delta K = m_0 c^2 (\gamma - 1) = 2.179784822\text{E-}18\,\text{j} \tag{2.11}$$

If we now divide this value by the unit charge of the electron, we again obtain a value in electronvolts equal to (13.6 eV).

So, we observe that both ΔK and Δm_m resolve to the same 13.6 eV energy value for this stated velocity, that we may be strongly tempted to consider as representing the same energy quantum calculated by different means.

But, it can hardly be disputed that on one hand, Δm_m measures the energy contained in a mass increment corresponding to an increase of the global magnetic field of the electron and that on the other hand, ΔK measures the well known kinetic energy that propels the effective mass of the electron at the stated velocity, an effective mass that includes by structure the quantity Δm_m calculated with Equation (2.2) on top of including the invariant rest mass of the electron.

Consequently, the only possible conclusion is that these two instances of 13.6 eV are different and are both induced simultaneously in the electron at this velocity, and consequently are in reality two *half-quanta* of energy whose sum constitutes a single quantum of *carrying energy* that exists separately from the energy quantum of which the electron invariant rest mass is made, one of which converting to a mass increment, while the other remains vectorially unidirectional, propelling *the total effective mass* of the electron at the stated velocity.

All calculations with Equations (2.2) and (2.6) for any velocity will reveal that this even split between an amount going into an increase in magnetic field mass and a translational momentum related amount of kinetic energy is maintained for the whole range of all possible relativistic velocities.

Interestingly, the total amount of 27.2 eV that results from adding the energy of the magnetic mass increment obtained from Equation (2.2) and the momentum energy obtained from Equation (2.6), is exactly equal to the single amount of energy that can be calculated with the Coulomb equation as a function of the mean axial distance separating the electron ground state orbital from the hydrogen nucleus, and that corresponds to relativistic reference velocity 2187647.561 m/s:

$$E = \int_{a_0}^{\infty} \frac{1}{4\pi\varepsilon_o} \frac{e^2}{a_0^{\,2}} \cdot da_0 = 0 - \frac{1}{4\pi\varepsilon_o} \frac{e^2}{a_0} = -4.359743805\,\text{E-}18\,\text{J} \tag{2.12}$$

When dividing this amount of energy by the unit charge value (1.602176462E-19 C), we effectively obtain in electronvolts the exact amount of energy obtained by summing up the energies obtained from Equations (2.2) and (2.6), that is, 27.2 eV, which confirms the validity of Equation (2.2) newly derived by Marmet, on top of confirming the fact that this total amount of energy induced for any relativistic velocity in a charged particle can be entirely obtained from an equation stemming from electromagnetism, that is, the Coulomb Equation (2.12), which now allows reuniting both ΔK and Δm_m obtained from Equations (2.2) and (2.11) as belonging to a single quantum of energy now directly related to electromagnetism since they are simultaneously induced by the Coulomb force. For example, the electron carrying energy at distance a_0=5.291772083E-11 m from the proton can be formulated as:

$$\text{Charged particle carrying energy} = \Delta K + \Delta m_m c^2 = 4.359743805\text{E} - 18\text{j} \tag{2.13}$$

2.4. The issue of momentum energy being considered conservative

Examining Equation (2.13) now reveals a major disconnect between the traditional classical/relativistic mechanics concept of *momentum*, that can be related only to the ΔK half of the energy adiabatically induced by the Coulomb force, which is deemed, from the traditional perspective, to reduce to zero when a body is not in motion, even if it remains adiabatically induced from the electromagnetic perspective, when the electron is captive in the hydrogen ground state orbital, in which it is now well understood that it is not moving on the theoretical Bohr orbit, as put in clear perspective in Reference ([43], [8] Chapter 2).

Moreover! There is no trace in traditional classical/relativistic mechanics, nor in traditional quantum mechanics, of the second component of Equation (2.13), that is, $\Delta m_m c^2$, which is adiabatically induced by the Coulomb force simultaneously with the ΔK component.

In classical/relativistic mechanics, momentum is obviously viewed as the most fundamental principle, a concept that was carried on to traditional quantum physics under the forms of the Hamiltonian and the Lagrangian. But in electromagnetism, the energy that sustains momentum is even more fundamental than momentum, given that it still remains adiabatically present by definition even when this momentum is inhibited, that is, even when an electrically charged particle, such as the electron, is stopped in its motion when captured in a state of axial electromagnetic equilibrium in one of the least action orbitals in an atom ([43], [8] Chapter 2).

This fundamental disconnect between electromagnetism on one hand, and traditional classical mechanic, traditional relativistic mechanics and traditional quantum mechanics on the other, makes it all the more difficult to conceptually overcome, since the value of ΔK as calculated with Equation (2.11) depends uniquely on the *velocity* parameter, which means that if this velocity falls to zero, then no momentum, that is, no motion inducing kinetic energy is conceptually deemed to exist from the non-electromagnetism traditional perspectives, which is in flagrant contradiction with the fact that according to Equation (2.13) stemming from electromagnetism, this energy is adiabatically induced uniquely as a function of the axial distance between electrically charged particles by the Coulomb force, that forbids by very nature that any other level of energy be induced between two charges separated by this distance, which means that it can only remain induced even if the velocity of the particle is inhibited, as demonstrated in Reference ([43], [8] Chapter 2). See Section 3.23.

Even from the Quantum Mechanics perspective, the wave function accounts for the complete physical presence of this 13.6 eV ΔK momentum energy via the Hamiltonian even if it is experimentally established that the electron is unable to move toward the proton at any velocity despite the impossibility by structure that this momentum energy be vectorially oriented in any direction other than toward the proton.

This observation consequently brings to light the possibility that momentum kinetic energy can exist as a *material substance* irrespective of whether or not a forward velocity is involved, as analyzed in detail in References ([15], [8] Chapter 6) ([43], [8] Chapter 2) ([36], See also Section 3.17), and is at the heart of a new paradigm that now allows mechanically explaining a series of electromagnetic processes that find no explanation from the traditional conservative principles perspective ([36], See also Chapter 3) ([34], [8] Chapter 19).

Having now made this connection, the analysis that follows will be carried on strictly from the electromagnetism perspective.

2.5. Separating the energy of the varying magnetic field increment from that of the invariant magnetic field of the rest mass of the electron

This new perspective now allows us to clearly separate the carrying energy of the electron from that of its rest mass and to calculate separately their electromagnetic frequencies and wavelengths with the standard equations $E=hv$ and $c=\lambda v$. We thus obtain the following frequency and electromagnetic wavelength for the 4.359743805E-18 j reference carrying energy of the electron on the theoretical Bohr ground state orbit, which in fact corresponds to the mean carrying energy of the electron in the ground state orbital of the hydrogen atom:

$$v = \frac{E}{h} = 6.579683909\text{E}15\,\text{Hz} \qquad \lambda = \frac{c}{v} = 4.55633521\text{E}-08\,\text{m} \qquad (2.14)$$

Similarly, we obtain the following frequency and electromagnetic wavelength for the energy of $E=m_oc^2=$ 8.18710414E-14 j making up the electron invariant rest mass, which wavelength is also known as the electron Compton wavelength:

$$v = \frac{E}{h} = 1.235589976\text{E}20\,\text{Hz} \qquad \lambda_C = \frac{c}{v} = 2.426310215\text{E}-12\,\text{m} \qquad (2.15)$$

We thus immediately observe that the energy related to the electron in motion involves the presence of not only one single simple harmonic electromagnetic oscillation, as the Schrödinger wave function seems to currently assume, but of two distinct harmonic oscillations, whose mutual resonance interaction is not yet clearly defined.

These values will be quite useful later on when the zitterbewegung motion of the moving electron will be analyzed in Section 2.18, as well as in Section 2.20, the complex electron resonance beat involving the interaction of these two harmonic oscillations with those of the proton inner electromagnetic elementary components when the electron is captive in the rest orbital of the hydrogen atom.

Let us also remark that although the concept of *wavelength* is sometimes assumed to represent a physical *length* to be associated to localized photons or even to the hypothetical electromagnetic waves of Maxwell's theory, such a wavelength can only be in reality a physical *distance* that the transversely oscillating electromagnetic energy half-quantum of such a localized photon or spread out theoretical electromagnetic wave needs to travel in space for one of the cycles of mutual induction of its transverse electric and magnetic aspects to be completed in reference to its frequency.

Speaking of Maxwell's continuous electromagnetic wave concept, the experiments carried out by Huygens, Fresnel and Young that demonstrate that at the macroscopic level, when a macroscopic electromagnetic wavefront is made to meet a surface into which a small aperture is made, however small it may be from our macroscopic perspective, this small aperture becomes the source of a secondary spherical electromagnetic wavefront, which is often considered as *the proof* of the physical existence of continuous electromagnetic waves such as Maxwell conceived them.

There is a habit in the community of thinking of an *electromagnetic wavefront*, but in reality there is an uninterrupted flow of electromagnetic energy in all of space, whether considered as a continuous wave phenomenon or as a crowd of countless separate point-like behaving electromagnetic photons that are constantly being individually emitted by de-exciting electrons in atoms, after these electrons were excited either out of atoms or just pushed further away from their nuclei to some metastable orbital.

In reality, this behavior of electromagnetic energy as measurable at our macroscopic level does not demonstrate any disconnect with the idea that this macroscopic electromagnetic wavefront could be made in reality of countless elementary point-like behaving electromagnetic photons that would interact, while moving through the small apertures, with the countless other point-like behaving electromagnetic elementary particles captive in various least action electromagnetic equilibrium states in the atoms making up the inner sides of the macroscopic apertures, and whose trajectories would consequently be deflected in such a way as to produce what seems to be, from our macroscopic perspective, *secondary spherical electromagnetic wavefronts* observed as they come out of the aperture.

There is absolutely nothing that rules out either the possibility that the individual photons emitted by de-exiting electrons in atoms all over the universe could continue behaving point-like after emission until they are subsequently absorbed by other charged particles, re-initiating in so doing the emission process, after having had their trajectories deflected numerous times, losing each time some energy as work in accordance with the 2nd Principle of thermodynamics with each resulting change in direction, before being absorbed by other charged particles at some other locations, as analyzed in Reference ([15], [8] Chapter 6).

Whether one concludes that electromagnetic energy really exists as a continuous wave phenomenon as perceived from our macroscopic level or that localized photons are the real thing at the submicroscopic level holds only to what a person has studied. Both schools of thought always had quite respectable adepts. The fact is that even if treating electromagnetic energy as localized quanta is consistent with the results of experiments carried out at the submicroscopic level, treating it as a continuous wave phenomenon remains consistent with the results of experiments carried out at our macroscopic level.

It seems however that the conclusion according to which this energy would physically exist as localized photons, as concluded by Planck, Einstein, de Broglie and Schrödinger, among others, allows clearer mechanical explanations of the various processes at the submicroscopic level.

2.6. Particularities of energy calculation by means of the Coulomb equation

Even though Equation (2.12) calculates the carrying energy of the electron at the hydrogen atom mean ground orbital distance from the central proton by mathematically accumulating this energy from *infinity* to this specific distance from $r=0$, it can be observed that this amount of energy can only be systematically equal to the actual amount of kinetic energy adiabatically induced by the Coulomb force as a function of this distance separating both electrical charges, a distance which is equal by structure to the distance separating point d from point *zero* in the integration function (**Figure 2.1**).

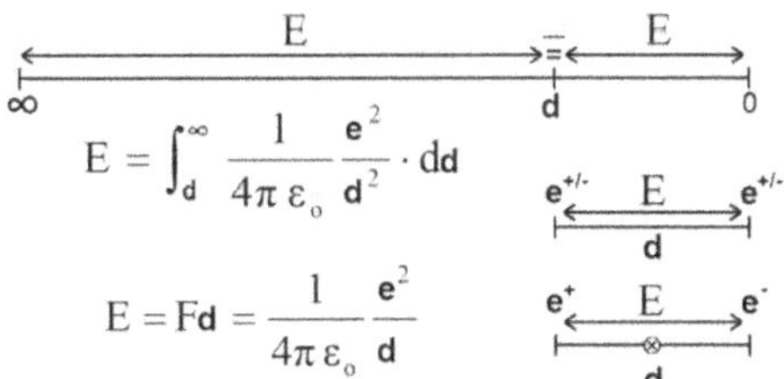

Figure 2.1: Energy equality between integration from infinity to distance d and between d and zero.

It can be observed also that point zero of the integration function can be relocalized to the middle of the distance separating both charges processed by the Coulomb equation without affecting energy calculation in any way, a central point $\otimes$ that will be later correlated with the central junction point of an expanded spatial geometry.

The method used by Marmet to derive Equation (2.9) from the Biot-Savart equation then allows deriving a new more general form of the Coulomb equation equivalent to traditional equation $E=h\nu$, that allows calculating the energy of any electromagnetic energy quantum without any need to use the Planck constant, and that also allows defining their intrinsic E and B fields strictly by means of a set of known electromagnetic constants.

Isolating the value of m_o in Equation (2.9) established by Marmet and making use of the familiar equation $\mu_0\,\varepsilon_0 c^2 = 1$ stemming from equivalent second partial derivatives of Maxwell's Equations ([14], [8] Chapter 13), that as far back as the 1860's allowed him to calculate the invariant speed of light from the two fundamental constants of vacuum ε_o and μ_o, will allow introducing the electrostatic permittivity constant of vacuum ε_o to link up with the Coulomb equation. Isolating μ_o in this equation in the following manner $\mu_0 = 1/\varepsilon_0 c^2$ allows replacing it with its equivalent electromagnetic definition ([30], [8] Chapter 4):

$$m_0 = \frac{\mu_0\, e^2}{4\pi\, r_e} = \frac{e^2}{4\pi\varepsilon_0\, r_0 c^2} \tag{2.16}$$

Multiplying then both left and right terms of Equation (2.16) by c^2 will convert the equation from calculating mass to calculating energy, in this particular case, the energy quantum of which the invariant rest mass of the electron is made:

$$E = m_0 c^2 = \frac{e^2}{4\pi\varepsilon_0\, r_0} = 8.18710414\,\mathrm{E}-14\ \mathrm{j} \tag{2.17}$$

Assuming that e^2 is likely to represent any pair of charges in such a general equation, let us replace the *classical electron radius r_o* used by Marmet by the classical radius of the Bohr theoretical orbit in the Bohr atom a_o to remain coherent with the hydrogen atom example. Considering that the Coulomb force between two such charges requires involving the distance between the charges, let us further divide both sides of the equation by a_o to finally obtain the equation allowing to calculate the Coulomb force, and to identify in the resulting equation the long established electrostatic constant whose exact value is 8.987551733E-9 N·m²/c², also know as the Coulomb constant:

$$F = \frac{E}{r_0} = \frac{e^2}{4\pi\varepsilon_0\, r_0^{\,2}} \quad \text{where} \quad \frac{1}{4\pi\varepsilon_0} = k_e \ \text{(Coulomb constant)} \qquad (2.18)$$

As a final validity confirmation, let us calculate the well-known Coulomb force that applies at the theoretical Bohr orbit, using the Bohr radius a_o=5.291772083E-11 m:

$$F = \frac{e^2}{4\pi\varepsilon_0\, a_0^{\,2}} = 8.238721807\text{E} - 08\,\text{N} \qquad (2.19)$$

Now going back to Equation (2.17) that allows calculating the energy making up the electron rest mass, we observe that the only *possibly variable* parameter that determines the amount of energy of the quantum making up this mass is r_o, which is considered a fundamental constant known under the name of *the classical electron radius* and that Marmet used to derive Equation (2.9).

It is well understood in physics circles that despite its name, this constant cannot really be an actual *radius* of the electron, given that it is now well established experimentally that the electron behaves *point-like* in all scattering experiments. *Point-like behavior* meaning here that during all such scattering experiments, however energetic, no unbreachable limit was ever detected at some distance from electrons' centers, however close two electrons came to each others' centers.

So despite this unfortunate misleading misnomer, r_o is nevertheless considered useful to define a *length*, or *distance*, not yet fully understood related to electromagnetic interactions involving electrons at the submicroscopic level.

But we may now have *telltales* as to what this *length* or *distance* may be, beginning with the observation that if we use it in Equation (2.12) instead of the classical Bohr radius a_o, we obtain the actual energy of the quantum of which the rest mass of the electron is made, as just calculated with the standard Equation (2.17), as if r_o was a really existing *distance* between a pair of still to be identified internal *would-be charges* involved in the still to be established electromagnetic inner oscillating structure of the electron, despite the fact that the *electric* charge of the electron is known to be unique and set at the fixed value of 1.602176462E-19 C. Maybe some sort of point-like *charges* of some nature different from *electric* but that would still be acted upon by the Coulomb force, despite the strangeness of the idea.

We will see further on that such an internal structure has effectively been established involving a harmonic oscillation of the inner magnetic energy of the electron cyclically converting to two such *non-electric charges* and back to magnetic energy state. See Equation (2.53) further on.

One more telltale relates to the relation between r_o and λ_c that is, the electron Compton wavelength, that we just calculated with Equation (2.15). This further telltale relates to a relation between these two *constants* and fine structure constant α, first described in Reference ([53], [8] Chapter 6) in relation with the recall constant of Hooke's law as applied to the LC transverse oscillation of the magnetic energy of the rest mass of the electron, and that amounts to half the rest mass of the electron, as determined by Marmet with Equation (2.9).

From the calculation carried out in Reference ([30], [8] Chapter 4), the maximum transverse separation amplitude between these *charges* during this reciprocating LC oscillation would be exactly equal to $r_o=\alpha\lambda_c/2\pi$, which would amount to the maximum distance that these two still to be identified *non-electric would-be charges* would reach in space as they oscillate between this double-component state and the magnetic single component constituting the magnetic half of the energy of the electron invariant rest mass. This conclusion was confirmed later when the LC oscillating *neutrinic charges* of the electron were identified in Reference ([33], [8] Chapter 12). This issue will be discussed further on.

This relation between r_o, λ_c and α led to consider the possibility that the same method could be applied to calculate the energy of any self-sustaining electromagnetic quantum, and subsequent verification confirmed the possibility. It thus turns out that $r=\alpha\lambda/2\pi$ coincides with the maximum distance that two charges -- either electric or neutrinic -- can reach transversely during the LC oscillation of any self-sustaining electromagnetic quantum during the reciprocating oscillation that causes them to cyclically induce the magnetic field of the particle as they close in on each other, and its regression as they move away from each other ([53], [8] Chapter 6) ([33], [8] Chapter 12) ([31], [8] Chapter 11) ([32], [8] Chapter 14) as we will see further on.

This means in fact that r_o and a_o are not really fundamental constants but only special cases of the whole range of possible electromagnetic energy transverse amplitudes coinciding with two quantized stationary action stable states of electromagnetic energy, that is, the invariant rest mass of the electron, and the stationary action electromagnetic equilibrium state of the electron in the hydrogen atom, and that they can be systematically replaced in the Coulomb equation by the more general variable expression $\alpha\lambda/2\pi$, λ being the electromagnetic longitudinal wavelength traditionally related to the electromagnetic energy quantum considered.

This allowed defining the following general equation in Reference ([30], [8] Chapter 4) by adapting Coulomb Equation (2.19) in the following manner (see also **Figure 2.1**):

$$E = \int_{a_0}^{\infty} \frac{1}{4\pi\,\varepsilon_o} \frac{e^2}{(\alpha\lambda/2\pi)^2} \cdot dr = 0 - \frac{1}{4\pi\,\varepsilon_o} \frac{e^2\,2\pi}{\alpha\lambda} = \frac{e^2}{2\,\varepsilon_o\alpha\lambda} \tag{2.20}$$

which defines an electromagnetic equation equivalent to $E=hv$, but that does not require the use of Planck's constant to calculate electromagnetic energy levels, whose complete derivation and justification was established in Reference ([30], [8] Chapter 4):

$$E = hv = \frac{e^2}{2\,\varepsilon_o\alpha\lambda} \tag{2.21}$$

A surprising benefit brought about by the establishment of this form of the Coulomb equation was that it finally allows unifying all classical force equations by allowing to reversibly derive the fundamental equation $F=ma$ from all of them ([44], [8] Chapter 7), beside observing that the Coulomb equation is an integral part of the Biot-Savart equation, since it is derived from Marmet's derivation from the Biot-Savart equation. See also Subsections 1.7.1 to 1.7.3.

2.7. Separately calculating the E and B fields of the electron and those of its carrying energy

The development of Equation (2.21) then allowed separately defining in Reference ([30], [8] Chapter 4) the E and B fields equations accounting for the totality of the energy of which the invariant rest mass of the electron is made:

$$\mathbf{B} = \frac{\mu_0\pi e c}{\alpha^3\lambda_C^{\,2}} = 8.289000222\text{E}13\,\text{T} \quad \text{and} \quad \mathbf{E} = \frac{\pi e}{\varepsilon_0\alpha^3\lambda_C^{\,2}} = 2.484979751\text{E}22\,\text{N/C} \tag{2.22}$$

and with the same equation, using the electromagnetic wavelength of its carrying energy, to calculate the E and B fields of this carrying energy. To remain consistent with our example of hydrogen atom ground state orbital, here are the E and B fields calculated with the wavelength of the carrying energy obtained in Equation (2.14):

$$\mathbf{B} = \frac{\mu_0 \pi e c}{\alpha^3 \lambda^2} = 235051.7341\,\text{T} \quad \text{and} \quad \mathbf{E} = \frac{\pi e}{\varepsilon_0 \alpha^3 \lambda^2} = 7.046673712\text{E}13\,\text{N/C} \tag{2.23}$$

Reference ([30], [8] Chapter 4) then shows how magnetic and electric fields Equations (2.22) and (2.23) can be added to establish the combined $\mathbf{E}$ and $\mathbf{B}$ fields of the electron in motion. To remain consistent with the hydrogen atom ground state parameters, the wavelengths obtained with Equations (2.14) and (2.15) are used to calculate the corresponding fields:

$$\mathbf{B} = \frac{\pi \mu_0 e c}{\alpha^3} \frac{\left(\lambda^2 + \lambda_c^2\right)}{\lambda^2 \lambda_c^2} = 8.289000246\,\text{E}13\,\text{T} \tag{2.24}$$

$$\mathbf{E} = \frac{\pi e}{\varepsilon_0 \alpha^3} \frac{\left(\lambda^2 + \lambda_c^2\right)\sqrt{\lambda_c(4\lambda + \lambda_c)}}{\lambda^2 \lambda_c^2 \,(2\lambda + \lambda_c)} = 1.813341121\text{E}13\,\text{N/C} \tag{2.25}$$

It can now be confirmed that Equations (2.24) and (2.25) are valid by calculating with the values obtained, the well known relativistic velocity of the electron when moving with the 4.359743805E-18 j reference energy of the hydrogen ground state (27.2 eV):

$$v = \frac{\mathbf{E}}{\mathbf{B}} = \frac{1.813341121\text{E}13}{8.289000246\text{E}13} 10^{-7} = 2,187,647.566\,\text{m/s} \tag{2.26}$$

The reason why the result must be multiplied by 10^{-7}, is that this factor, which was made part of the definitions of ε_0 and μ_0 for these constants to remain in harmony with the CGS system when the MKS units were adopted ([14], [8] Chapter 13), and that are part of the parameters required to calculate the $\mathbf{E}$ and $\mathbf{B}$ fields of the moving electron with Equations (2.24) and (2.25), ends up being squared in the denominator of the $\mathbf{E/B}$ fraction of Equation (2.26), which is something not obvious unless actual calculations are carried out, as in our example. This unwanted squaring is circumvented by simply multiplying the equation by 10^{-7} during its resolution. See Reference ([14], [8] Chapter 13) for an explanation of why this factor must not be squared.

So we observe that the magnetic mass increment provided by Marmet's Equation ([29], Equation (M-17), which was previously reproduced as Equation (2.2), can be matched with a corresponding $\mathbf{B}$ field provided by Equation (2.23) from the electromagnetic wavelength of 4.556335256 E-8 m of the corresponding quantum of energy 4.359743805 E-18 j, thus amending Equation (2.10) to obtain the mass increment using the velocity stemming from the $\mathbf{E}$ and $\mathbf{B}$ fields as calculated with Equation (2.26).

Given that Equation (2.26) provides the same relativistic velocity that Marmet established from the gamma factor [29] from Equations (2.4) and (2.5), and that he used in establishing Equation (2.2), the velocity term of Marmet's equation can be replaced by the ***E/B*** relation that defines this velocity in Equation (2.26):

$$\Delta m_m = \frac{\mu_0 \left(e^-\right)^2}{8\pi r_e}\frac{v^2}{c^2} = \frac{\mu_0 \left(e^-\right)^2}{8\pi r_e}\frac{(\mathbf{E}/\mathbf{B})^2}{c^2} = 2.425337726\mathrm{E}-35\,\mathrm{kg} \tag{2.27}$$

thus allowing for the first time the calculation of a *classical mass* strictly from electromagnetic parameters, with no need to involve any varying velocity parameter.

The magnetic energy density involved can now be established for the energy of the composite ***B*** field calculated with Equation (2.24):

$$u_B = \frac{\mathbf{B}^2}{2\mu_0} = \frac{1}{2\mu_0}\left(\frac{\pi\mu_0 ec}{\alpha^3\lambda^2\lambda_C{}^2}\right)^2\left(\lambda^2 + \lambda_C{}^2\right)^2 = 2.733785555\,9\mathrm{E}33\ \mathrm{j/m}^3 \tag{2.28}$$

For comparison, here is the density of the magnetic field of the isolated invariant rest mass of the electron, making use of the invariant magnetic field of the electron calculated with Equation (2.22):

$$u_B = \frac{\mathbf{B}^2}{2\mu_0} = \frac{1}{2\mu_0}\left(\frac{\mu_0\pi ec}{\alpha^3\lambda_C{}^2}\right)^2 = 2.733785544\mathrm{E}33\ \mathrm{j/m}^3 \tag{2.29}$$

and that of the isolated carrying energy of the electron in the hydrogen ground state calculated with Equation (2.23) is:

$$u_B = \frac{\mathbf{B}^2}{2\mu_0} = \frac{1}{2\mu_0}\left(\frac{\mu_0\pi ec}{\alpha^3\lambda^2}\right)^2 = 2.198300502\mathrm{E}16\ \mathrm{j/m}^3 \tag{2.30}$$

The equation defining the volume within which such high energy densities make sense is derived in Reference ([30], [8] Chapter 4) and is also shown further on as Equation (2.50).

2.8. *The internal electromagnetic structure of the electron carrying energy*

It has been established long ago that the electron is an electromagnetic particle. However, the nature of its carrying energy could never be clarified until Marmet derived Equation (2.2) from the Biot-Savart equation, leading to Equation (2.13), that reveals that this carrying energy is made of 2 parts, that is, one half sustaining the momentum ΔK of the particle, and the other half identified by Marmet as an amount of magnetic energy that adds a relativistic mass increment Δm_m to the invariant rest mass of the particle in motion.

Since the electric charge of the electron has been systematically proven over the course of the past century to remain invariant, irrespective of its velocity, the related intrinsic electric E field that was established with the second Equation (2.22) can be expected to also remain invariant, and to remain Maxwell equations compliant, so does its intrinsic magnetic B field established with the first Equation (2.22).

As put in perspective with Equation (2.13), since the Δm_m magnetic mass increment identified by Marmet increases in the exact same proportion as the ΔK momentum energy of the electron, and that these two energy amounts can not be part of the energy quantum making up the invariant rest mass of the electron, this gives us a first conclusive clue that this carrying energy also is electromagnetic in nature, since its magnetic field cannot be dissociated from electromagnetism and consequently from Maxwell's equations.

This total amount of carrying energy represented by Equation (2.13) can thus be logically represented with the following relational equation:

$$E_{\left(\substack{Total\ electron \\ carrying\ energy}\right)} = E_{\left(\substack{Momentum \\ energy}\right)} + E_{\left(\substack{Magnetic\ mass \\ increment\ energy}\right)} \tag{2.31}$$

But to remain consistent with electromagnetism, it seems impossible that this magnetic energy component would not be involved in some cyclic process of electromagnetic oscillation between this magnetic state and a yet to be identified *electric* state, that could potentially be represented as a reciprocating oscillation between both states, in conformity with the very foundation of Maxwell's theory, to the effect that for electromagnetic energy to even exist, both of these aspects must mutually induce each other [17]:

$$E_{\left(\substack{Total \\ carrying\ energy}\right)} = E_{\left(\substack{Momentum \\ energy}\right)} + \left[E_{\left(\substack{Electric \\ state}\right)} \cos^2(\omega t) + E_{\left(\substack{Magnetic \\ state}\right)} \sin^2(\omega t) \right] \tag{2.32}$$

It is at this point that a huge leap *out of the box* needs to be made, as the saying goes, because since this magnetic mass increment freshly identified by Marmet has been proven to physically exist by transverse interaction with relativistically moving electrons in experiments carried out by Walter Kaufmann at the beginning of the 20th century [36], this means that the energy making up this *mass increment* can only physically exist just like the energy making up the invariant rest mass of the electron. And finally so must it also be for its momentum energy, despite the long held conclusion that it exists only inasmuch as its velocity can be expressed.

This conclusion leads to converting relational Equation (2.32) to the following electromagnetic form, representing this electromagnetic oscillation as a simple harmonic LC reciprocal *transverse* oscillation – in conformity with the fact that the $\boldsymbol{E}$ and $\boldsymbol{B}$ fields must be perpendicular to the direction of motion – between an electric state and a magnetic state of the energy making up the magnetic mass increment identified by Marmet:

$$E_{\left(\substack{Total\ carr\ ying \\ energy}\right)} = \frac{hc}{2\lambda} + \left[\frac{e^2}{2C_\lambda}\cos^2(\omega t) + \frac{L_\lambda\, i_\lambda^{\,2}}{2}\sin^2(\omega t) \right] \tag{2.33}$$

where

$$E_{E(\max)} = \frac{e^2}{2C} \quad \text{and} \quad E_{B(\max)} = \frac{L\, i^2}{2} \tag{2.34}$$

The definitions of the subcomponents C, L and i are provided further on with Equations (2.45) and (2.47).

In this transitory form, Equation (2.33) may give the impression that the electromagnetic energy of the Δm_m half-quantum oscillates *longitudinally*, so to speak, moving in the same vectorial direction as its $\Delta K=hc/2\lambda$ momentum energy, but we will see further on that it can oscillate only transversely in conformity with Maxwell's equations, when the vectorial infrastructure will be put in place with Equation (2.48).

We will also see further on that the oscillation of the magnetic energy of this magnetic mass increment, between a state of maximum presence and one of zero presence as a function of its electromagnetic frequency is key to understanding the various resonance states of the electron, that is its zitterbewegung motion on one hand, and also its axial resonance state when captive into a least action electromagnetic equilibrium state in an authorized atomic orbital. See Sections 2.18 and 2.20.

Indeed, it can be established, as we will see further on, that even the magnetic energy of the invariant rest mass of the electron can only be involved separately in a similar harmonic oscillating motion between maximum presence and zero presence in space ([31], [8] Chapter 11), and that the same oscillation characterizes the magnetic energy of the two types of elementary components making up all nucleons and of their respective carrying energies, that is, the up quark and the down quark ([32], [8] Chapter 14).

2.9. *Correlating classical mechanics and relativistic mechanics via electromagnetism*

The first benefit of representing the electron carrying energy with LC Equation (2.33), is the ease with which it allows visualizing its electromagnetically oscillating half as oscillating perpendicularly to the direction of motion of the energy sustaining its translational momentum $\Delta K=hc/2\lambda$, which clearly matches, as previously mentioned, the well known perpendicular relation between the E and B fields of Maxwell's theory with respect to the direction of motion of any point on the wave front of his theoretical continuous electromagnetic wave in spherical expansion from its point of emission.

In turn, this clear separation between the unidirectionally oriented momentum energy and the transversely oscillating energy of the carrying energy quantum allowed directly upgrading Newton's non-relativistic kinetic energy equation $K=mv^2/2$ to a fully relativistic electromagnetic form ([42], [8] Chapter 5):

$$\frac{v^2}{c^2} = \frac{4\lambda\lambda_C + \lambda_C^{\ 2}}{\left(2\lambda + \lambda_C\right)^2} \tag{2.35}$$

One unexpected outcome of the establishment of Equation (2.35) was that when using the wavelength of the carrying energy induced at mean ground state orbital distance from the nucleus of the hydrogen atom 4.556335261E-08 m, it directly provides the fine structure constant α ([60], [8] Chapter 8):

$$\alpha = \frac{v}{c} = \frac{\sqrt{\lambda_C\left(4\lambda + \lambda_C\right)}}{\left(2\lambda + \lambda_C\right)} = 7.29735253\ 3\mathrm{E}-03 \tag{2.36}$$

More surprising yet! Further dividing Equation (2.36) by 2π provides the exact fine structure constant α related electron g factor discovered by Julian Schwinger in 1948 ([60], [8] Chapter 8) [70]:

$$\begin{pmatrix}\text{Electron} \\ \text{magnetic momen} \\ \text{magnetic drift}\end{pmatrix} = \frac{\sqrt{\lambda_C\left(4\lambda + \lambda_C\right)}}{2\pi\left(2\lambda + \lambda_C\right)} = \frac{\delta\mu}{\mu_B} = \frac{\alpha}{2} = 1.161386535\mathrm{E}-3 \tag{2.37}$$

The fact that electromagnetic Equation (2.35) is relativistic by structure, also allows deriving the 4 standard relativistic equations. First in line is the relativistic energy equation, now amended to also account for the presence of the magnetic mass increment Δm_m of elementary particles carrying energy ([42], [8] Chapter 5):

$$K = 2m_0 c^2 (\gamma - 1) \tag{2.38}$$

For the first time ever also, apparently, the Lorentz gamma factor equation was derived directly from an electromagnetic equation in Reference ([42], [8] Chapter 5), that is, from Equation (2.35), instead of from strictly mathematical geometric and trigonometric methods, as has been systematically done since Woldemar Voigt came up with the idea in 1887 ([36], See also Section 3.4) ([42], [8] Chapter 5) ([60], [8] Chapter 8) [71] [72]:

$$\gamma = \frac{1}{\sqrt{1 - v^2/c^2}} \tag{2.39}$$

The third relativistic equation derived was of course the relativistic mass equation of a moving elementary particle in motion ([42], [8] Chapter 5):

$$E = \gamma mc^2 \quad \text{where} \quad \gamma m = m_o + \Delta m_m \tag{2.40}$$

And finally, the relativistic energy-momentum relation Equation (See Appendix A):

$$E^2 = (pc)^2 + (mc^2)^2 \tag{2.41}$$

Thus conclusively demonstrating that classical relativistic equations and electromagnetic equations can reversibly be derived from each other.

Besides Equation (2.35) making use of the wavelengths defined with Equations (2.14) and (2.15) from which all classical relativistic equations can be derived, a second and even more fundamental electromagnetic equation was derived from upgrading Newton's kinetic energy equation to full electromagnetic status ([42], [8] Chapter 5). It is the following equation that directly makes use of the *quantities of energy* separately making up the invariant rest mass of the electron, its momentum, and finally its magnetic mass increment, the last two constituting its carrying energy. It is the following form:

$$\frac{(hc/\lambda + 2hc/\lambda_c)^2 - (2hc/\lambda_c)^2}{\left((2L_c\ i_c^2) + (L_\lambda\ i_\lambda^2)\right)^2} = \frac{v^2}{c^2} \tag{2.42}$$

that resolves to

$$v = c\frac{\sqrt{4EK_{momentum} + (K_{momentum})^2}}{2E + K_{magnetic}} \tag{2.43}$$

where E represents the energy of the invariant rest mass of the electron, $K_{momentum}$ is the ΔK momentum energy provided by the carrying energy, and

$K_{magnetic}$ is the energy going into the Δm_m magnetic mass increment provided by the carrying energy of the electron.

What is so fundamental and important about this equation, is that when the energy of the electron rest mass is reduced to zero, leaving only its carrying energy in the equation, we end up with an equation that systematically provides the speed of light in an invariant manner, whatever the sum of the two half-quanta always equal by structure of the momentum energy and of the energy of the remaining magnetic mass; a velocity possible only for free moving electromagnetic energy:

$$v = c\,\frac{K_{momentum}}{K_{electromagnetic}} = \frac{\Delta K}{\Delta m_m c^2} = c\,\frac{(hc/2\lambda)}{(L_\lambda\ i_\lambda^{\,2})} = c\,\frac{1}{1} = 299{,}792{,}458\,\text{m/s} \qquad (2.44)$$

where

$$L = \frac{\mu_0 \alpha \lambda}{8\pi^2} \quad \text{and} \quad i = \frac{2\pi\,ec}{\alpha\lambda} \qquad (2.45)$$

Since Marmet's contribution allows conclusively establishing that Δm_m from Equation (2.2) and ΔK from Equation (2.6) will systematically be equal whatever total amount the sum of their energies will be, these two energy values will systematically simplify to 1 in Equation (2.44) whatever energy amount of electromagnetic energy is represented by its wavelength λ.

This means that for the first time, we have a conclusive clue regarding the possible internal electromagnetic structure of localized free moving electromagnetic photons, that is, electromagnetic photons that would not be slowed down by having to *carry and propel*, so to speak, the translationally inert electromagnetic mass of an electron on top of having to carry and propel their own electromagnetic mass complement. So LC Equation (2.33) could consequently be applied as well to free moving electromagnetic photons and to the electron's carrying energy, which would fully justify naming the latter a "*carrier-photon*".

2.10. The de Broglie double-particle electromagnetic photon

So let us now consequently re-identify Equation (2.33) as describing the total energy of a free-moving electromagnetic photon and analyze further its structure:

$$E_{\left(\substack{Total\ photon \\ energy}\right)} = \frac{hc}{2\lambda} + \left[\frac{e^2}{2C_\lambda}\cos^2(\omega t) + \frac{L_\lambda\,i_\lambda^2}{2}\sin^2(\omega t) \right] \qquad (2.46)$$

Of course, the L and i variables definitions of Equations (2.45) still apply and the definition of C established in Reference ([15], [8] Chapter 6) is:

$$C = 2\varepsilon_0 \alpha \lambda \qquad (2.47)$$

We first observe that the electric phase of the electromagnetic transverse oscillation between magnetic and electric states seems to involve a pair of charges, which is a possibility that has been a major stumbling block in electromagnetic theory ever since Maxwell established his theory of light propagation on the then axiomatic concept that the very existence of this energy mandated that both E and B fields mutually induce each other for the energy to even exist.

Even if the resulting theory has proven out of any doubt its absolute conformity with experience at the macroscopic level, the origin of the *displacement current* that would involve such a local motion of some postulated double electric charges to induce the magnetic field, while they supposedly close in on each other, inducing the magnetic field, to then be re-induced themselves as the magnetic field regresses, could never be clarified either experimentally nor theoretically.

In a search to identify these still hypothetical charges at the submicroscopic level, de Broglie tried in the 1930's to establish a clear internal electromagnetic mechanics of the localized photon grounded on the characteristics of the wave function.

It turns out that he did correctly establish that such a permanently localized photon should satisfy the Bose-Einstein's statistic and Planck's Law, explain the photoelectric effect while obeying Maxwell's equations and remain in accordance with the properties of Dirac's theory of complementary corpuscles symmetry, if it involves two half-photons of spin 1/2,

> *"... qui doivent être complémentaires l'un de l'autre dans le même sens que l'électron positif* [le positon] *est complémentaire de l'électron négatif dans la théorie des trous de Dirac... Un tel couple de particules complémentaires est susceptible de s'annihiler au contact de la matière en cédant toute son énergie, ce qui rend compte parfaitement des caractéristiques de l'effet photoélectrique... le photon étant constitué de deux particules*

élémentaires de spin h/4π, il doit obéir à la statistique de Bose-Einstein comme l'exige l'exactitude de la loi de Planck pour le rayonnement noir... ce modèle du photon permet de définir un champ électromagnétique lié à la probabilité d'annihilation du photon, champ qui obéit aux équations de Maxwell et possède tous les caractères de l'onde électromagnétique lumineuse." ([27], p.277).

Translation:

"... that must be complementary with respect to each other in the same manner that the positive electron [the positron] *is complementary to the negative electron in the Dirac Hole Theory... Such a complementary couple of particles is likely to annihilate at the contact of matter by relinquishing all of its energy, which perfectly accounts for the characteristics of the photoelectric effect... The photon, being made up of two elementary particles of spin h/4π, will obey the Bose-Einstein statistic as required by the precision of Planck's law for the black body....this model of the photon allows the definition of an electromagnetic field linked to the probability of annihilation of the photon, a field that obeys Maxwell's equations and has all the characteristics of electromagnetic light waves."*

His attempts to define the localized electromagnetic photon from the properties of the wave function were unsuccessful to the point that he finally concluded in 1936 that it was impossible to exactly represent elementary particles in the frame of 4D spacetime geometry, in his view too restrictive, hinting that if this frame could eventually be escaped from, such a description could become possible:

"... la non-individualité des particules, le principe d'exclusion et l'énergie d'échange sont trois mystères intimement reliés : ils se rattachent tous trois à l'impossibilité de représenter exactement les entités physiques élémentaires dans le cadre de l'espace continu à trois dimensions (ou plus généralement de l'espace-temps continu à quatre dimensions). Peut-être un jour, en nous évadant hors de ce cadre, parviendrons-nous à mieux pénétrer le sens, encore bien obscur aujourd'hui, de ces grands principes directeurs de la nouvelle physique." ([27], p. 273).

Translation:

Retrospectively, it seems that in the restricted frame of 4D space-time geometry, reverse engineering the description of the electromagnetic photon from the characteristics of the wave function that was not initially grounded on electromagnetism to start with, was an impossible task, since, let's remember, it was introduced by Schrödinger to represent a state of resonance in the classical resonance mechanics sense, stemming from de Broglie's comparison with well known classical mechanics resonance states [58]. See also Equation (2.1). We will come back further on to this issue of reverse engineering in Section 2.19. See also Section 1.2 on this issue.

The only real relation that can exist between Schrödinger's wave function and the *electromagnetic* resonance state of the electron captive in least action electromagnetic resonance state in the ground state orbital of the hydrogen atom can then only be a description of the spatial resonance volume within which all of the electron energy is expected to be contained, and gives no clue whatsoever as to the nature of the *electromagnetic resonator* whose resonance characteristics could explain this resonance volume.

Besides, the very idea that half the energy of the quantum could behave as 2 half-quantities displaying *electric* properties as they close in on each other while at the same time concentrically accumulating as a single quantity within the same volume of space that would display *magnetic* properties goes directly against logic when considering that this energy would be a *physically existing substance* as the previous analysis leads to conclude, which would imply that it interpenetrates itself as it oscillates.

This mechanical impossibility that comes to light when attempting to represent in the same volume of space the mutual induction of the electric and magnetic aspects of localized electromagnetic quanta by reciprocating swing, effectively correlates with de Broglie's conclusion that elementary particles cannot be represented in the too restrictive frame of 4D spacetime geometry.

2.11. Expanding the space geometry

In Maxwell's wave theory, it is well understood that the continuous wave concept imposes that both E and B fields of Maxwell's theory must be *in phase* for the wave to exist and propagate. But contrariwise, the idea that the energy of localized electromagnetic quanta could exist, due to a self-sustaining reciprocating LC oscillation, imposes that both fields be 180° *out of phase* for such an LC oscillation to be mechanically possible.

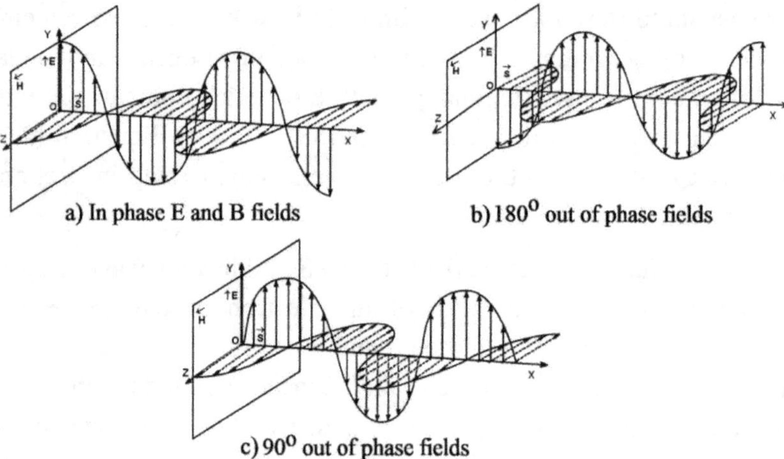

Figure 2.2: Traditional representations of electromagnetic fields in phase, 180° out of phase, and 90° out of phase, in classical electromagnetism.

A close examination of the traditional graphic representations of Maxwell's theory electromagnetic phases and of his equations reveals however that both in phase alignment and 180° out of phase alignment result in the very same configuration (**Figure 2.2**).

This reveals that although 180° out of phase alignment is incompatible with Maxwell's continuous electromagnetic wave theory, it is perfectly allowed by his equations, and that true 180° out of phase alignment, involving that electric energy reaching minimum while the magnetic energy reaches maximum and the reverse is indeed allowed and is effectively consistent with representing a self-sustaining electromagnetic quantum by a reciprocating LC oscillating process (**Figure 2.3**). Moreover, it is consistent with the very foundation of Maxwell's theory to the effect that both fields have to mutually induce each other for the energy to even exist.

With regard to the mechanical impossibility that 2 half-quantities of a physically existing *substance* displaying *electric* properties as they close in on each other while at the same time concentrically accumulating in a single quantity that would display *magnetic* properties within the same volume of space, it is this very mechanical impossibility that gave rise to the idea that the solution could be for the magnetic quantity to *grow*, so to speak, into a different space while both charges were closing in within the first space, and inversely.

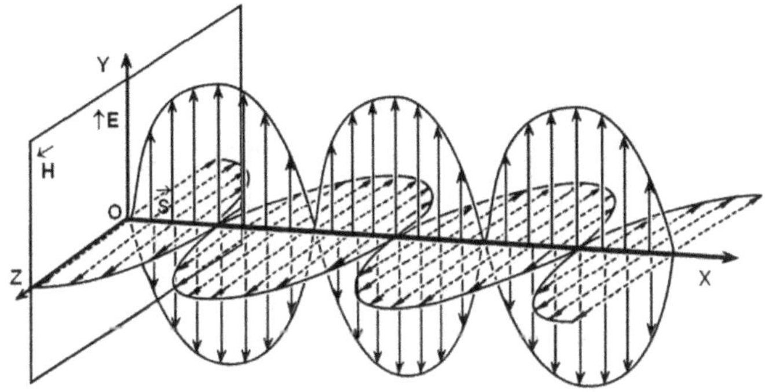

Figure 2.3. 180° out of phase representation of **E** and **B** fields of Maxwell's theory for an LC oscillation.

And without even going so far as to assume the real physical existence of such a second space, it so happens that from the vectorial perspective, it is rather easy to represent such a multi-spaces complex, and it is particularly easy to represent vectorially both E and B fields of the Δm_m magnetic mass half-quantum as transversely oscillating with respect to the direction of motion of the ΔK momentum half-quantum, in conformity with Maxwell's equations.

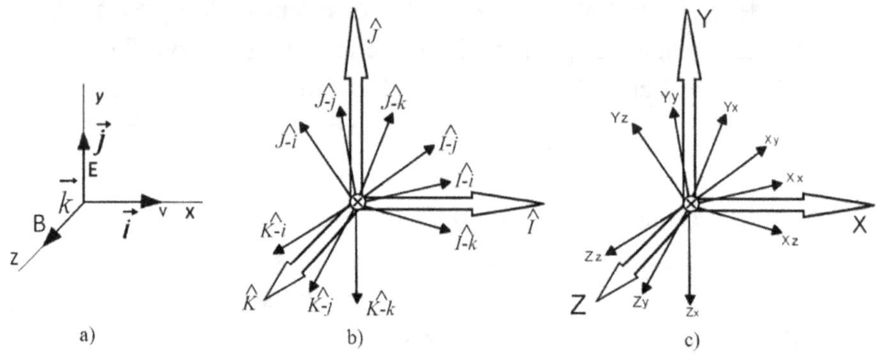

Figure 2.4: Major and minor unit vectors set applicable to the trispatial geometry.

In this particular case, it so happens that the well known vectorial cross product of the magnetic *B* field vector and electric *E* field vector, both perpendicular to each other, resolving in a third vector perpendicular to the first two and representing the phase velocity (**Figure 2.4-a**), which is the triple orthogonal relation that maps the direction of motion at the speed of light of any point of the wavefront of Maxwell's hypothetical spherically expanding continuous electromagnetic wave, gives us a solid footing to explore such a possibility.

The method consists in geometrically *exploding*, so to speak, each of the 3 standard electromagnetic vectors *i*, *j* and *k*, applicable to normal space into 3 full blown 3D vectorial spaces of their own (**Figure 2.4-b**), each of the three X, Y and Z spaces (**Figure 2.4-c**) remaining perpendicular to the other two and all of them remaining connected via their common origin, previously identified as being midpoint ⊗ between a pair of charges in **Figure 2.1**, and that can now be seen as a transit point for the energy, that would be located at the center of each electromagnetic elementary quantum, and through which the *substance* of the energy quantum would be free to travel as if between communicating vessels, according to their required electromagnetic reciprocating motion, without the illogical interpenetration of the *energy substance* that would prevent such a reciprocating motion within the more limited frame of a single 3D space geometry.

Contrary to expectation, it turns out to be relatively easy to mentally visualize such a trispatial mutually orthogonal 9-dimensional geometric complex. It suffices to imagine each of the 3 sets of minor vectors *i, j* et *k* of **Figure 2.4-b** as if they were folded metaphoric 3-rib umbrellas.

This allows us to mentally open at will any one of them, one at a time, to full 3-axes orthogonal extension to observe the behavior of the energy quantum substance in this fully deployed 3D space during each phase of the oscillating motion. **Figures 2.4-b** and **2.4-c** show the dimensions of the 3 spaces half deployed to allow clear unique identification of each of the 9 resulting orthogonal inner axes, which allows simple mathematical and vectorial identification of the inner motion of the energy within each space without changing nor invalidating in any way any of the traditional vectorial representations applied in the normal 4D space geometry to represent electromagnetic energy, in the traditional mechanics.

In this space geometry, the momentum energy that translationally propels elementary particles is unidirectional by definition, and is set by structure to be insensitive to any transverse interaction, which directly correlates with the observations made by Walter Kaufmann between the longitudinal inertia and the transverse inertia of electrons moving at relativistic velocities in a bubble chamber [36], when he observed that although both half-quanta ΔK and Δm_m could be measured longitudinally in addition to the electron rest mass, only the Δm_m half-quantum could be measured transversely in addition to the electron rest mass.

The same property will cause the pair of opposite signs *electric charges* of an electromagnetic quantum unidirectionally moving toward or away from each other on the Y-y/Y-z plane within Y-space to appear to be neutral with respect to the orthogonally oriented Y-x axis and to not even be detectable as perceived from X-space, which is the space from which we observe objective reality, which correlates with the observed fact that electromagnetic photons do not seem to possess electric charges ([15], [8] Chapter 6) ([53], [8] Chapter 6), despite the physical incompatibility of such an absence with Maxwell's theory.

The same indetectability and apparent absence of opposite signs charges will characterize the pairs of *neutrinic charges* unidirectionally moving toward or away from each other on the X-y/X-z plane within X-space ([33], [8] Chapter 12) ([31], [8] Chapter 11).

The fact that the pair of *electric charges* can only move in opposite directions on the Y-y/Y-z plane is what explains why photons can be polarized at any angle perpendicularly to their direction of motion along the X-x axis of normal X-space. Obviously, the same polarizability property will apply to the pair of *neutrinic charges* moving in opposite directions on the X-y/X-z plane.

Finally, any amount of energy now oscillating between Y and Z spaces finds itself oscillating transversely *by structure* with respect to normal X-space, and will thus appear to possess omnidirectional inertia as perceived from X-space, that is, will behave as if it was *massive* in the sense understood in classical/relativistic mechanics, as perceived from the X-space.

This expanded space geometry was first proposed at the Congress-2000 event held at the St Petersburg State University in July of 2000 [28]. It is introduced and put in perspective in Reference ([34], [8] Chapter 19) with respect to the traditional multidimensional geometries conceived of in previous historical attempts at resolving the remaining issues of fundamental physics, and is completely described in Reference ([15], [8] Chapter 6).

2.12. Fundamental symmetry maintained by structure

One aspect of utmost interest of the trispatial geometry is that the fundamental principle of symmetry is respected by structure for all aspects of the distribution of the energy of an electromagnetic quantum.

The energy is systematically distributed between one half remaining unidirectional in one of the spaces while the other half cyclically oscillates in perpendicular orientation with respect to the first half by structure (*a half-half symmetry*), which is what immediately reveals that in this space geometry, the speed of light can only be an invariant equilibrium velocity in vacuum in all cases of free moving electromagnetic photons, due to this half-half energy distribution mandated by structure between both half-quanta ([15], [8] Chapter 6).

Within electrostatic Y-space where both electric charges – for free moving photons and carrier-photons – axially oscillate on the Y-y/Y-z plane toward and away from each other ([15], [8] Chapter 6) ([53], [8] Chapter 6), and within normal X-space where both neutrinic charges (for massive electron, positron, up quark and down quark - speaking only of the stable states) ([33], [8] Chapter 12) ([31], [8] Chapter 11) ([32], [8] Chapter 14) also axially oscillate, but on the X-y/X-z plane toward and away from each other in the same manner on this plane perpendicular to Y-space in which their unidirectional complement resides along the Y-x axis, always symmetrically possessing equal amounts of energy and opposite directions, along which the varying distance between them, provide the corresponding varying intensity of the opposite signs of their charges (symmetry between the energy amounts and also between the intensities of the opposite signs of their charges within Y-space and X-space).

Within magnetostatic Z-space where a single quantity of the energy grows to a maximum while leaving Y-space for photons and carrier-photons ([15], [8] Chapter 6) ([53], [8] Chapter 6), or while leaving X-space for massive particles ([33], [8] Chapter 12) ([31], [8] Chapter 11) ([32], [8] Chapter 14), this single quantity, after having reached maximum volume presence in Z-space, regresses toward zero presence in this space while the energy crosses over back into the Y-space – or X-space – it was in previously (symmetry between the increasing phase and the decreasing phase of the energy presence within magnetostatic Z-space). See Section 2.15.

In normal X-space, neutrino energy can be released only as identical pairs in opposite directions perpendicularly to the direction of motion of the unidirectional energy present in this space belonging to a newly created massive elementary particle – electron, muon or tau – that sheds in this manner an excess metastable initial excess amount of mass ([33], [8] Chapter 12) (More on this issue later on in this Chapter).

And global symmetry is also preserved since the time-varying space-wise moving electric dipole is permanently counterbalanced by a related time-varying growing and decreasing magnetic dipole oriented perpendicularly, with both dipoles remaining perpendicular to the direction of motion of the photon in space, thus obeying the triple orthogonality required for plane wave treatment in Maxwell's theory's for straight line motion of electromagnetic energy ([15], [8] Chapter 6).

2.13. The trispatial photon equation

The first inner electromagnetic structure that the trispatial geometry allowed defining was that of the localized photon that de Broglie concluded could not be defined within the too restrictive confines of 3D space ([15], [8] Chapter 6), and that graphically shows with **Figure 2.5** the transverse harmonic oscillation sequence of the energy of the photon as represented with Equation (2.46).

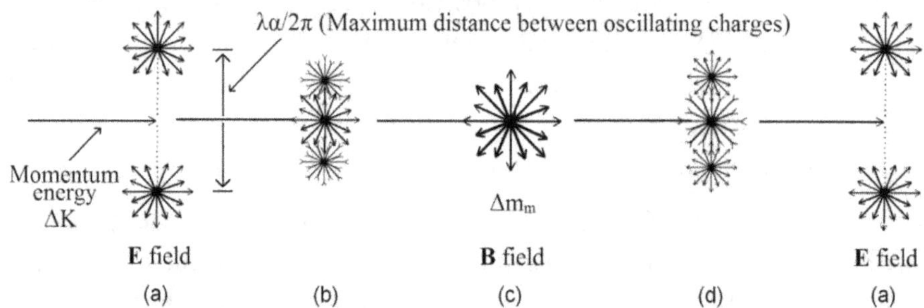

Figure 2.5. Complete time varying cycle of the energy in transverse oscillation of the electromagnetic half-quantum of the double-particle photon while its momentum sustaining unidirectional half-quantum propels it translationally.

Figure 2.5 indeed allows visually representing the complete time varying sequence of the transverse oscillation of the energy of the electromagnetic half-quantum within the trispatial complex. **Figure 2.5-a** shows both inner opposite charges, measurable as generating the *E* field of the photon at its maximum value, having reached maximum transverse distance within electrostatic Y-space, followed by **Figure 2.5-b** showing the energy of both charges transferring to magnetostatic Z-space.

Then comes **Figure 2.5-c** showing the energy of both charges having completely entered Z-space in omnidirectional expansion, now measurable as generating the *B* field of the photon at its maximum value, followed by **Figure 2.5-d** showing the energy of the single magnetic component transferring back into electrostatic Y-space. Finally, the final **Figure 2.5-a** shows all of the magnetic energy having transferred back into Y-space as two opposite charges having reached again maximum transverse distance, and again measurable as the *E* field of the photon, ready for the next cycle to begin.

As mentioned previously, the double-particle photon concept is an original idea of Louis de Broglie, and the complete analysis of its elaboration in the trispatial geometry is available in Reference ([15], [8] Chapter 6), where the complete development of its trispatial LC equation is elaborated from the inductance and capacitance representations of electromagnetic energy:

$$E\,\vec{I}\,\vec{i} = \left(\frac{hc}{2\lambda}\right)_{X}\vec{I}\,\vec{i} + \left[\begin{array}{l} 2\left(\dfrac{e^2}{4C}\right)_{Y}(\,\vec{J}\,\vec{j},\vec{J}\,\overleftarrow{j}\,)\cos^2(\omega t)\\[2ex] +\left(\dfrac{L\,i^2}{2}\right)_{Z}\overleftrightarrow{K}\ \sin^2(\omega t) \end{array}\right] \qquad (2.48)$$

and also of the same LC formulation making use of the more familiar E and B fields defined with Equations (2.23):

$$E\,\vec{I}\,\vec{i} = \left(\frac{hc}{2\lambda}\right)_{X}\vec{I}\,\vec{i} + \left[\begin{array}{l} 2\left(\dfrac{\varepsilon_0\mathbf{E}^2}{4}\right)_{Y}(\,\vec{J}\,\vec{j},\vec{J}\,\overleftarrow{j}\,)\cos^2(\omega t)\\[2ex] +\left(\dfrac{\mathbf{B}^2}{2\mu_0}\right)_{Z}\overleftrightarrow{K}\ \sin^2(\omega t) \end{array}\right]V \qquad (2.49)$$

Where volume V is the theoretical stationary isotropic volume that the incompressible oscillating kinetic energy of the photon would occupy if it was immobilized as a sphere of isotropic density, as derived in Reference ([30], [8] Chapter 4):

$$V = \frac{\alpha^5}{2\pi^2}\lambda^3 \qquad (2.50)$$

2.14. The trispatial electron equation

It is well established that electromagnetic photons of 1.022 MeV or more can be destabilized into converting to an electron-positron pair ([31], [8] Chapter 11). However, it so happens that all of the energy making up the two 0.511 MeV/c^2 rest masses of both electron and positron is electromagnetic in nature and thus reside within Y-space and Z-space in the new trispatial geometry, while the half-quantum of the whole initial quantum of a 1.022 MeV photon that resides in X-space before decoupling is vectorially unidirectional by definition. This means that Nature has found some way to force this ΔK unidirectional momentum energy to re-orient transversely for it to become part of the electromagnetic mass of both emergent massive particles.

One of the most interesting features of the trispatial geometry is that it effectively allows establishing a clear mechanical process by which this ΔK unidirectional energy of the half-quantum sustaining the momentum of a moving electromagnetic photon of 1.022 MeV can cross over into the orthogonal electrostatic Y-space and magnetostatic Z-space during the decoupling process, and so acquire the transverse orientation property that characterizes the complete mass energy of both electron and positron of the pair resulting from the separation process in the trispatial geometry ([31], [8] Chapter 11).

By the same token, the very mechanics of transfer of this momentum energy into Y-space to end up defining the invariant unit charge of both electron and positron, also forces by structure the other half of the energy of each particle of the pair in process of separation to now start oscillating between Z-space and X-space for the symmetric energy distribution be maintained in the trispatial complex, resulting in the establishment of a double component state within X-space separating in a manner identical to the behavior of the pair of *electric charges* of the photon within Y-space, that are traditionally represented by e^2, but that now require to be identified with a new denomination since they cannot have the *electric* characteristic anymore, that belongs by definition only to energy present in Y-space, in this trispatial complex. Pending clear identification, the first-draft symbol that best fitted them was then $(e')^2$.

As we will see further on, deeper analysis succeeded in relating these double $(e')^2$ *non-electric charges* to the emission of neutrinos, which earned them the name of *neutrinic charges* in the subsequent descriptions ([33], [8] Chapter 12) ([31], [8] Chapter 11).

The following trispatial LC equations were then defined to describe the corresponding inner trispatial energy structure of the mass of the electron and positron:

$$\vec{E}\vec{0}=m_ec^2\vec{0}=\left[\frac{H}{2\lambda_C}\right]_Y\vec{J}\vec{i}+\left(\begin{array}{l}2\left[\dfrac{(e')^2}{4C_C}\right]_X(\vec{I}\vec{j},\vec{I}\overleftarrow{j}\,)\cos^2(\omega t)\\[2em]+\left[\dfrac{L_Ci_C^{\,2}}{2}\right]_Z\overleftrightarrow{K}\,\sin^2(\omega t)\end{array}\right) \qquad (2.51)$$

and

$$\vec{E}\vec{0}=m_pc^2\vec{0}=\left[\frac{H}{2\lambda_C}\right]_Y\vec{J}\overleftarrow{i}+\left(\begin{array}{l}2\left[\dfrac{(e')^2}{4C_C}\right]_X(\vec{I}\vec{j},\vec{I}\overleftarrow{j}\,)\cos^2(\omega t)\\[2em]+\left[\dfrac{L_Ci_C^{\,2}}{2}\right]_Z\overleftrightarrow{K}\,\sin^2(\omega t)\end{array}\right) \qquad (2.52)$$

Subsequent reformulation of the same LC equations making use of the more familiar E and B fields defined with Equations (2.23) then forced the identification of the component pair $(e')^2$ as *neutrinic charges* (v^2) in References

([33], [8] Chapter 12) ([31], [8] Chapter 11) for reasons that will soon become obvious:

$$m_0 \vec{0} = \frac{V_m}{c^2} \left\{ \left[\frac{\varepsilon_0 E^2}{2} \right]_Y \vec{J}\,\vec{i} + \left[2\left(\frac{\varepsilon_0 \boldsymbol{v}^2}{4} \right)_X (\vec{I}\,\vec{j}, \vec{I}\,\overleftarrow{j})\cos^2(\omega t) + \left(\frac{\boldsymbol{B}^2}{2\mu_0} \right)_Z \overleftrightarrow{K}\sin^2(\omega t) \right] \right\} \tag{2.53}$$

where v (Greek letter **nu**) represent the *neutrinic* field Equation ([31], [8] Chapter 11) ([33], [8] Chapter 12), that now represents the double *neutrinic charges* whose energy calculation is identical to that of the electric E field equation, but that now oscillate in opposite directions on the X-y/X-z plane of X-space, within the trispatial energy structure of elementary massive particle masses in Equations (2.51) and (2.52), just like the *electric charges* oscillate in opposite directions on the Y-y/Y-z plane of Y-space, within the photon or carrier-photon trispatial energy structure ([15], [8] Chapter 6) ([53], [8] Chapter 6) in Equation (2.48) and (2.49). Here are the definitions of the required isotropic volume and of the neutrinic field:

$$V_m = \frac{\alpha^5 \lambda_C^{\,3}}{2\pi^2} \quad \text{and} \quad V = \frac{\pi(e')}{\varepsilon_0 \alpha^3 \lambda_C^{\,2}} \tag{2.54}$$

where $(e')^2$ is assigned the same numerical value grounded on the value of the unit electric charge $e=1.602176462E\text{-}19$, since a pair of such components represents by structure the same maximum amount of energy in the trispatial electron structure, that is, half the rest mass of the electron when reaching maximum distance from each other within X-space as it splits in two equal quantities.

Modelled on Equation (2.49) for the free moving photon, the equation for the electron carrier-photon making use of E and B fields can now be formulated as follows, providing the same energy as the corrected classical relativistic kinetic Equation (2.13):

$$E_K \vec{I}\,\vec{i} = \left[\frac{hc}{2\lambda} \right]_X \vec{I}\,\vec{i} + \left[2\left(\frac{\varepsilon_0 E_K^{\,2}}{4} \right)_Y (\vec{J}\,\vec{j}, \vec{J}\,\overleftarrow{j})\cos^2(\omega t) + \left(\frac{B_K^{\,2}}{2\mu_0} \right)_Z \overleftrightarrow{K}\sin^2(\omega t) \right] V_K \tag{2.55}$$

which now allows representing the combined fields equations of the electron and its carrying energy as in **Table 2.1**.

In fact, the carrier-photon provides the electron with the ambient E and B fields that permanently determine its velocity and direction of motion, when they can be expressed, in accordance with the Lorentz equation $F = q(E + v \times B)$ previously mentioned. More precisely, it constantly obeys the triple orthogonal relation $v=E/B$ stemming from the Lorentz equation imposed to it by the E and B fields of its carrier-photon, whose intensity determines its velocity, and whose relative densities equilibrium determine its trajectory, default equal densities of both E and B fields resulting in straight line motion of the electron ([30], [8] Chapter 4).

Table 2.1: Combined fields equations of the moving electron and its carrier-photon

	Momentum kinetic energy in X-Space (normal space)	Energy located in Y and Z spaces making up the translationally inert mass of the particle in motion
Rest mass energy (m_oc^2)		$\left\{\left(\dfrac{\varepsilon_0\mathbf{E}^2}{2}\right)_Y \vec{J}\,\vec{i}+\left(\left(\dfrac{\mathbf{B}^2}{2\mu_0}\right)_Z \overset{\leftrightarrow}{K}\right)\right\}V_{m_e}$
Carrying energy $\varDelta K + \varDelta m_m c^2$	$\left[\dfrac{hc}{2\lambda}\right]_X \vec{I}\,\vec{i}$	$\left[\left(\dfrac{\mathbf{B}_K{}^2}{2\mu_0}\right)_Z \overset{\leftrightarrow}{K}\right]V_K$
Total Relativistic mass energy (mc^2)		$\left\{\left(\dfrac{\varepsilon_0\mathbf{E}^2}{2}\right)_Y \vec{J}\,\vec{i}+\left(\left(\dfrac{\mathbf{B}^2}{2\mu_0}\right)_Z \overset{\leftrightarrow}{K}\right)\right\}V_{m_e}+\left[\left(\dfrac{\mathbf{B}_K{}^2}{2\mu_0}\right)_Z \overset{\leftrightarrow}{K}\right]V_K$

In turn, the B field of the electron carrier-photon constantly tends to align its relative magnetic polarity orientation, that is, its relative spin orientation, in least action antiparallel alignment with respect to the B field of the rest mass energy of the electron that it carries, and the combined resultant of which constantly tends to align in a least action antiparallel orientation with respect to the resultant of the B fields of the surrounding particles, thus with respect to the ambient macroscopic B field resulting from the addition of the surrounding B fields.

Given that the half-quantum ΔK momentum energy of the carrier-photon is immovably oriented perpendicularly to the B field of its own complementary electromagnetic mass increment Δm_m, the direction of motion of this unidirectional momentum energy is systematically determined by the orientation of its B field.

It is this immovable orthogonal relation that explains why unpaired electrons in ferromagnetic materials can be forced to align their spins parallel to each other in as best fit least action antiparallel mutual orientation as possible with respect to an ambient macroscopic magnetic B field, which forces their individual momentum ΔK energies to align in the same direction and add up to cause a macroscopic object such as the cylinder of the Einstein-de Haas experiment to rotate ([52], [8] Chapter 10), or reciprocally, this is why when the unidirectional ΔK momentum energies of the carrier-photons of unpaired electrons in the ferromagnetic rod of the Barnett experiment are forced to align parallel to each other by mechanically forcing the rod to rotate, their individual B fields are also forced to align in parallel spin orientation, and add up to become measurable at our macroscopic level ([52], [8] Chapter 10).

2.15. Neutrino emission in the trispatial geometry

Interestingly, the trispatial geometry allows establishing for the first time a mechanical explanation to the emission of neutrinos. This particular solution emerges from the mandatory LC structure of elementary electromagnetic quanta in the trispatial geometry.

From this perspective, given that it is well verified that the electric charge of newly created particles mu and tau remains invariant at the same unit value as that of the electron, it can be concluded from the viewpoint provided by the trispatial geometry that the energy corresponding to the metastable excess mass that these two particles display cannot enter Y-space, because any energy increase in this space would cause by structure an increase in the value of their electric charge, which we know experimentally never to occur.

Given that they are massive just like the electron, they have the same LC structure as the electron in the trispatial geometry. This involves that this excess energy can exist only as a metastable increase of the energy quantum that oscillates between Z-space and X-space. In retrospect, the same conclusion can be hypothesized for an electron newly created by β- decay, which would modify the trispatial LC rest mass Equation (2.53) of the electron in the following manner. To simplify equations representations, we will do away from now on with the now well established unit vectors notation:

$$m_{0+} = \left\{ \left[\frac{\varepsilon_0 \mathbf{E}^2}{2} \right]_Y + \left[\begin{array}{c} 2\left(\dfrac{\varepsilon_0 \left(\mathbf{V}_e + \mathbf{v}'\right)^2}{4} \right)_X \cos^2(\omega t) \\[2ex] + \left(\dfrac{(\mathbf{B}_e + \mathbf{B}')^2}{2\mu_0} \right)_Z \sin^2(\omega t) \end{array} \right] \right\} \frac{V_m}{c^2} \qquad (2.56)$$

where m_{o+} represents a slightly increased rest mass of the electron, and v' and B' are the slight energy increment that now momentarily oscillates between normal X-space and magnetostatic Z-space in momentary metastable excess energy with respect to the normal stable electron rest mass energy. This solution allows the electron electric field E to remain unchanged in conformity with observation.

Since this β- decay electron possesses slightly more energy than the well known invariant rest mass the electron, it seems quite possible that as it is in the process of leaving the destabilized neutron structure, the extreme destabilizing tensions due this initial proximity could force the electron's two neutrinic energy quanta into a violent translational motion about the X-x axis on the X-y/X-z plane, that would free the two excess half-quantities momentarily in excess, causing them to escape into normal X-space in opposite directions on this X-y/X-z plane perpendicularly to the direction of motion of the electron, while the two rest energy neutrinic quantities of the oscillating half-quantum of the electron recover their usual to and fro oscillation inside the inner electron structure, that now has reached its lowest possible energy level an henceforth invariant rest mass as represented with Equation (2.53).

$$m_{0+} \rightarrow m_0 + v_e + \overline{v}_e \qquad (2.57)$$

In the trispatial geometry, the muonic and tauic neutrinos emission would obviously follow the same pattern:

$$m_{0+} = \mu^- = \left\{ \left[\frac{\varepsilon_0 \mathbf{E}_e^{\,2}}{2} \right]_Y + \left[\begin{array}{c} 2\left(\dfrac{\varepsilon_0 \left(\mathbf{V}_e + \mathbf{V}_\mu\right)^2}{4} \right)_X \cos^2(\omega t) \\[2ex] + \left(\dfrac{(\mathbf{B}_e + \mathbf{B}_\mu)^2}{2\mu_0} \right)_Z \sin^2(\omega t) \end{array} \right] \right\} \frac{V_m}{c^2} \qquad (2.58)$$

$$m_{0+} = \tau^- = \left\{ \left[\frac{\varepsilon_0 \mathbf{E}_e{}^2}{2} \right]_Y + \left[\begin{array}{l} 2\left(\dfrac{\varepsilon_0 (\mathbf{V}_e + \mathbf{V}_\tau)^2}{4} \right)_X \cos^2(\omega t) \\[2ex] + \left(\dfrac{(\mathbf{B}_{e+} \mathbf{B}_\tau)^2}{2\mu_0} \right)_Z \sin^2(\omega t) \end{array} \right] \right\} \frac{V_m}{c^2} \qquad (2.59)$$

Resulting in similar mu and tau neutrinos emissions:

$$m_{0+} = \mu^- \rightarrow m_0 + v_\mu + \overline{v}_\mu \quad \text{and} \quad m_{0+} = \tau^- \rightarrow m_0 + v_\tau + \overline{v}_\tau \qquad (2.60)$$

Of course, β+ decay, anti-muon and anti-tau will result in identical neutrino emissions, leaving behind a single positron instead of an electron.

The fact that both neutrinos produced during each emission can only be released as an identical pair moving in opposite directions perpendicularly to the direction of motion of the emitting particle, makes it impossible for neutrinos produced by decaying muons coming in direct line from the Sun's surface in the general direction of the detector to be detected since they escape and move on planes perpendicular to the Sun-detector axis.

So according to the trispatial characteristics of neutrino emission, the only neutrinos/antineutrinos that can possibly be detected originating from the Sun will be a small fraction of those released by decaying muons in motion on a plane perpendicular to the Sun-detector axis, that is, mainly neutrinos emitted at the outside limits of the visible disk of the Sun, which is a conclusion that would go a long way in explaining why their detection rate has consistently remained so far below what current theories predict.

This conclusion could easily be verified experimentally by focusing detection equipment directly to the circumference of the solar disk.

Finally, since they escape as simple momentum related unidirectional kinetic energy amounts in X-space, deprived of the complementary transverse electromagnetic component oscillating between Y-space and Z-space that explains omnidirectional inertia as perceived from normal X-space, that is *electromagnetic mass*, as well as *electric charge*, for all electromagnetic elementary particles in the trispatial geometry, this provides a clear explanation of why no mass nor charges were ever detected for them in all experiments where they were involved.

2.16. Up and down quarks in the trispatial geometry

The last particles that must be examined before resonance states can be addressed are the up and the down quarks that have been confirmed to be the only scatterable point-like behaving electromagnetic charged massive elementary subcomponents that can be identified in protons and neutrons, during extensive non-destructive scattering experiments carried out from 1966 to 1968 at the SLAC facility ([34], [8] Chapter 19) ([32], [8] Chapter 14) [21].

The trispatial mechanics of creation of protons and neutrons from the only two possible combinations of triads of electrons and positrons interacting in close enough proximity without sufficient momentum energy to escape each others' mutual capture is described in Reference ([32], [8] Chapter 14).

Given that up and down quarks always display the same point-like behavior as electrons and positrons during all such scattering experiments with electrons or positrons, it has long been suspected in the community that these up and down quarks making up the inner scatterable structure of nucleons could possibly be positrons and electrons whose masses and charge characteristics would be warped into these potentially altered states by the stresses imposed by these most energetic least action equilibrium states that these particles can possibly reach in Nature ([34], [8] Chapter 19) ([32], [8] Chapter 14) ([43], [8] Chapter 2).

This possibility immediately brings to light a possible explanation to the observed fact that no up or down quark could ever be observed moving separately in space after having been scattered out of a nucleon by sufficiently energetic scattering. Indeed, it they really are electrons and positrons whose characteristics are warped into those observed for up and down quarks by their intensely stressed nucleonic electromagnetic environments, they would of course immediately recover their normal electron or positron characteristics as soon as they escape these warping stresses.

The specific electron and positron characteristics that would be warped out of skew by these intense stresses are first of all their masses, which have been determined to lie between 1 and 5 MeV/c^2 for the up quark and between 3 and 9 MeV/c^2 for the down quark, and their electric charges that have been determined to be 2/3 of the charge of the positron for the up quark, and 1/3 the charge of the electron for the down quark ([73], p. 382).

It so happens that the trispatial geometry allows defining a clear mechanics of creation of nucleons from the only two possible combinations of triads of electrons and positrons, which provides a logical explanation to these stress induced changes in characteristics, and also on the nature of these electromagnetic stresses ([32], [8] Chapter 14).

In the trispatial geometry, the mass and charge of stable elementary particles vary as an inverse function of each other as a function of their distance from the coplanar Y-z axis within electrostatic Y-space ([34], [8] Chapter 19) ([32], [8] Chapter 14).

The distance from the Y-z axis within Y-space at which an electron-positron pair decouples from a destabilized 1.022 MeV photon is by structure 3.861592641E-13 m ([31], [8] Chapter 11), which corresponds to the electron *classical radius* divided by the fine structure constant ($r'=r_e/\alpha$).

Table 2.2: Relation between up and down quarks charges and masses with respect to their distance from the Y-z axis within electrostatic Y-space.

Table of the effective charges and masses of the electron, the up quark and the down quark, estimated on the assumption that the unit charge of the electron would be the amount of charge induced at the distance from the Y-z axis at which electron-positron pairs separate during the pair production process.			
Particle	$r' = r_e/\alpha$	Charge	mass
Electron	r'_e = 3.861592641E-13 m	1.602176462E-19 C	9.10938188E-31 kg
Up quark	r'_{eu} = 2.574395094E-13 m	1.068117641E-19 C	2.04961092E-30 kg
Down quark	r'_{ed} = 1.287197547E-13 m	5.340588207E-20 C	8.19844378E-30 kg

At this distance from the Y-z axis, its charge exactly corresponds to the well known unit charge of 1.602176462E019 C and mass of 9.10938188E-31 kg. These values well established experimentally allow determining the corresponding values for the up and down quarks as in **Table 2.2** ([32], [8] Chapter 14), which fall precisely within the experimentally estimated limits for these masses.

This allows establishing the following general equation to calculate the invariant effective masses of the only three stable massive and electrically charged elementary electromagnetic particles, which are behaving point-like in all scattering encounters, that are the only electromagnetic elementary sub-components of all atoms that exist in the universe, by means of *the electrostatic energy induction constant*: K=1.220852596E-38 j·m^2 established from the Coulomb equation in References ([34], [8] Chapter 19) ([31], [8] Chapter 11) ([32], [8] Chapter 14) (See also Section 1.25). Of course, the positron can be considered as being identical to the electron except for the sign of its charge.

$$m_{i[d,u,e]} = K\left(\frac{3\alpha}{nr_0c}\right)^2 \qquad (n = 1,2,3) \tag{2.61}$$

In the trispatial geometry, this lessening of the charges of the up and down quarks due to the electromagnetic stresses they are subjected to inside nucleons cannot occur however without being compensated by an increase of the magnetic field of the particle and of its carrier-photon, as is demonstrated for the magnetic drift suffered by the electron carrier-photon energy, even as far from the nucleus at the ground state of the hydrogen atom ([60], [8] Chapter 8).

Given that the up and down quarks stabilize at such precise relative distances from the Y-z axis in the trispatial geometry, it becomes possible to establish their magnetic drift constants from these distances:

$$S_U = \frac{r'_{eu}}{r'_e} = \frac{2}{3} \qquad \text{and} \qquad S_D = \frac{r'_{ed}}{r'_e} = \frac{1}{3} \tag{2.62}$$

Table 2.3: Energies and wavelengths of the rest masses of the up and down quarks.

<table>
<tr><td colspan="4">Table of the energies and wavelengths of the effective masses of the up and down quarks, estimated on the assumption that the unit charge of the electron would be the amount of charge induced at the distance from the Y-z axis at which electron-positron pairs separate during the pair production process.</td></tr>
<tr><td>Particle</td><td>$r' = r_e/\alpha$</td><td>$E = K/r^2$</td><td>$\lambda = hc/E$</td></tr>
<tr><td>Electron</td><td>r'_e = 3.861592641E-13 m</td><td>0.5109989027 MeV</td><td>2.426310215E-12 m</td></tr>
<tr><td>Up quark</td><td>r'_{eu} = 2.574395094E-13 m</td><td>1.149747531 MeV</td><td>1.078360096E-12m</td></tr>
<tr><td>Down quark</td><td>r'_{ed} = 1.287197547E-13 m</td><td>4.598990173 MeV</td><td>2.69590021E-13 m</td></tr>
</table>

These magnetic drift constants and wavelengths now allow establishing the trispatial LC equations of both up and down quarks:

$$m_U = \frac{E_U}{c^2} = \frac{1}{c^2}\left\{ \left[S_U\left[\frac{hc}{2\lambda_U}\right]\right]_Y + (2-S_U)\left[2\left(\frac{(e')^2}{4C_U}\right)_X \cos^2(\omega t) + \left(\frac{L_U i_U^{\ 2}}{2}\right)_Z \sin^2(\omega t)\right]\right\} \qquad (2.63)$$

$$m_D = \frac{E_D}{c^2} = \frac{1}{c^2}\left\{ \left[S_D\left[\frac{hc}{2\lambda_D}\right]\right]_Y + (2-S_D)\left[2\left(\frac{(e')^2}{4C_D}\right)_X \cos^2(\omega t) + \left(\frac{L_D i_D^{\ 2}}{2}\right)_Z \sin^2(\omega t)\right]\right\} \qquad (2.64)$$

The carrier-photon of each up and down quark within nucleons would of course have the same internal trispatial LC structure as that of the electron carrier-photon, that is, that shown with Equation (2.55), and would relate to its carried particle in the same manner, as described in **Table 2.1** for the electron in motion, the only differences being the immensely higher energy levels that these nucleonic carrier-photons reach, and the amount of magnetic drift that they themselves suffer from the stress imposed by the nucleonic electromagnetic environment ([32], [8] Chapter 14).

This relation between each up quark and each down quark with each its own carrier-photon makes them amenable to being represented by a wave function in a manner similar to that of the electron in the hydrogen ground state, as we will see further on.

2.17. Parallel and anti-parallel relative magnetic spin orientations

In Quantum Mechanics (QM), the concept of *spin* is so weakly related to the magnetic field that although it is technically associated to the magnetic moment of charged particles, even this magnetic moment is seen by most in the community as a simple mechanical angular momentum ($Sz=\pm\frac{1}{2}\hbar$) with no clear reminder that it very specifically concerns the relative magnetic polarities orientation, either parallel or anti-parallel, of the magnetic fields of elementary electromagnetic quanta relative to each other. For all practical purposes, it is perceived as a mechanical *spinning motion* in two possible transverse directions perpendicularly to the direction of motion in the classical mechanics sense with no real relation with electromagnetism.

However, the very idea of a *magnetic spin* of elementary particles being akin to the classical/relativistic mechanics concept of *angular momentum* directly clashes with the experimentally confirmed fact previously mentioned that no unbreachable limit was ever detected at any distance from electrons' centers, however close two electrons came to each others' centers during absolutely all scattering encounters, because the very idea of an *angular momentum* implies the existence of a volume that can rotate, which is meaningless in the case of an elementary electromagnetic quantum such as the electron, for which no volume can be measured since it systematically behaves point-like in all scattering encounters.

The disconnect between the QM concept of *spin* and the physical relative polar magnetic orientations of elementary electromagnetic quanta is so important that many remain convinced that *spin* would be an *intrinsic* angular momentum property of particles, instead of what it can only be, that is a *relative* property that remains meaningless unless at least two electromagnetic quanta are involved, which is the uncircumventable condition for the very ideas of *magnetic parallel orientation* and *magnetic antiparallel orientation* to make any sense.

The fact that two electrons succeed so easily in associating in a very strong and intimate least action antiparallel magnetic spin covalent bounding to unite two hydrogen atoms into an H_2 molecule despite their electric repulsion function of the inverse square of the distance, *de facto* reveals that an interaction law of a higher order than the inverse square Coulomb force is simultaneously at play to so easily initiate and maintain such a powerful least action magnetic covalent bound between two electrons.

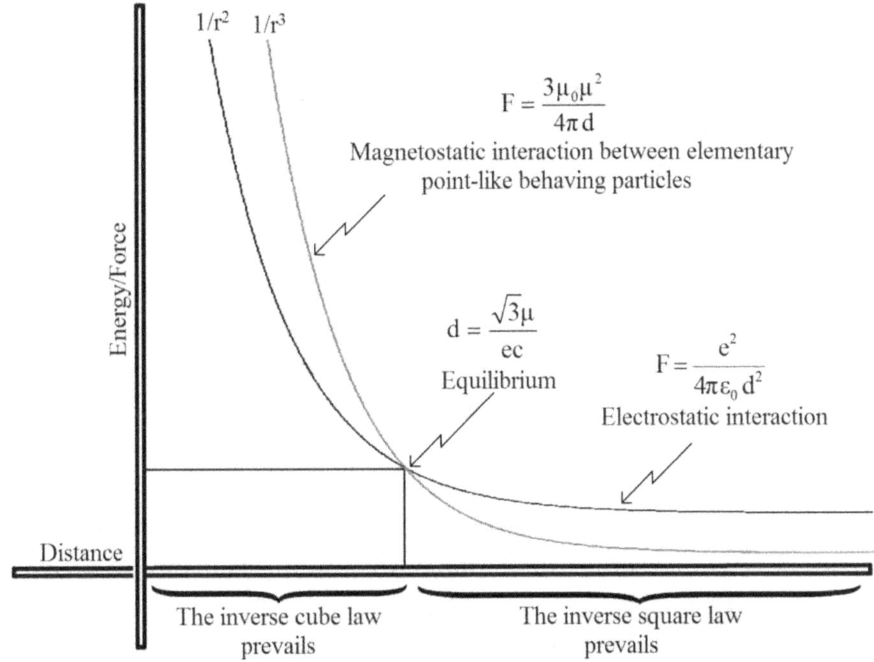

Figure 2.6: Intersecting inverse square and inverse cube interaction curves.

Interestingly, experiments carried out as recently as 2014 by Kotler et al. [64] experimentally demonstrated that the interaction law involved, when 2 electrons are forced to interact in parallel magnetic spin alignment, is the inverse cube interaction law function of the distance, which is the interaction that overcomes the inverse square repulsion Coulomb law when two electrons are forced to come sufficiently close to each other in antiparallel magnetic spin alignment. The relation between these two interaction laws is described in **Figure 2.6**.

Also, an experiment carried out in 1998 already confirmed this magnetic interaction function of the inverse cube between magnets having the same magnetic field configuration as that of elementary electromagnetic quanta, which allowed analyzing this magnetic interaction law in relation with the oscillating nature of the magnetic energy of electromagnetic quanta revealed in the trispatial geometry ([49], [8] Chapter 9), that brought to light the fact that the magnetic fields of elementary electromagnetic quanta behave at any given moment like magnetic monopoles that constantly reverse their polarity as a function of time according to their energy frequency ([11], See also Chapter 3).

This conclusion finally brings to attention the key function of the relative frequency ratios existing between elementary electromagnetic quanta in explaining (Section 3.20) why two electrons can so easily magnetically stabilize in covalent bounding despite their repelling same electric charges signs, due the ratio 1 of the synchronous frequencies of the spherical expansion and regression of their respective magnetic energies presence; also why an electron and a positron captive in metastable positronium configuration can combine to convert to electromagnetic photon states precisely due to the ratio 1 of their synchronous magnetic reversal frequencies combined with their attracting opposite electric charges signs [11]; and finally why an electron and a proton can so systematically magnetically repel each other to end up stabilizing at the known electron ground state orbital mean distance despite their attracting opposite electric charges signs ([11], See also Chapter 3) ([49], [8] Chapter 9), due to the asynchronous frequencies ratio of the spherical expansion and regression of their magnetic energy presence, whose mechanics was summarily analyzed in References ([43], [8] Chapter 2) ([11], See also Chapter 3) ([49], [8] Chapter 9), and that will be analyzed in more details further on in relation with the resulting resonance states.

But let us first analyze how the asynchronous magnetic interaction between the invariant frequency of the energy of the electron rest mass and the variable frequency of the energy of its carrier-photon allow both quanta to define the irregular resonance state known under the name of zitterbewegung of the electron in motion.

2.18 Zitterbewegung

Considering **Table 2.1** again that puts in perspective the fact that the electron in motion involves two different energy quanta, that not only electromagnetically oscillate at different frequencies, but whose harmonic oscillation centers ⊗ are physically separated by structure on a plane transverse with respect to the direction of motion of the system in space (See **Figure 2.7**).

Comparing the electron rest mass Equation (2.53) with its carrier-photon Equation (2.55) indeed shows that each quantum possesses its own trispatial junction ⊗ which are separated by structure from the simple fact that their energy oscillates between different pairs of spaces in the trispatial complex, that of the electron oscillating between Z-space and X-space, while that of its carrier-photon oscillates between Z-space and Y-space, on top of oscillating at different frequencies. This means that except for the case when the carrier-photon would possess exactly 0.511 MeV of energy, both components of the electron in motion are physically unable to associate in exactly synchronized attractive relative anti-parallel magnetic spin alignment, which highlights the contrast between these predictable and measurable asynchronous resonance interactions, and the unpredictable spontaneous stochastic fluctuations of the zero point energy level of QFT currently assumed to be responsible for zitterbewegung motion.

In reality, any difference in frequency between both components can only force both trispatial junctions ⊗ to follow oscillating trajectories that seems in appearance erratic transversely with respect to the direction of motion of the twin component system, due to the uninterrupted asynchronous sequence of cyclic alternance between attractive anti-parallel spin alignment states and repulsive parallel spin alignment states, which can only generate the resonance state that came to be identified as the zitterbewegung motion of the moving electron.

We will see further on that a third oscillation process, axial within atomic structures this time, will become involved when the electron is captured in least action electromagnetic equilibrium in atomic orbitals, which generates the actual three-components complex resonance volume within which de Broglie concluded the electron has to be captive into in the hydrogen atom and that Schrödinger meant to describe with the wave function.

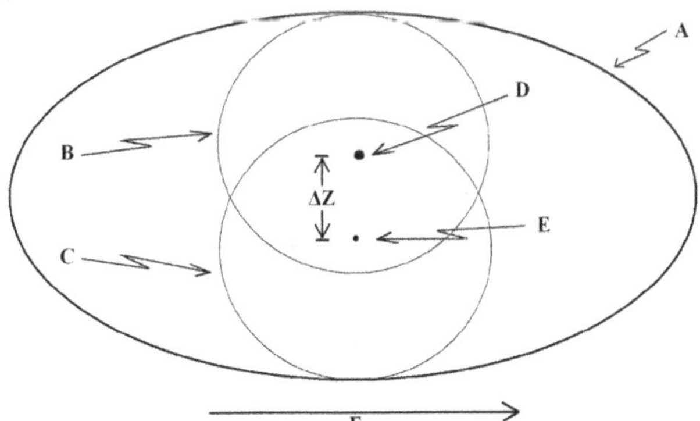

Figure 2.7: Free moving electron

Legends for **Figure 2.7**:

A - Symbolic representation of the resonance volume of the electron energy in free motion as definable by a wave function involving the cyclic magnetic spins reversal interaction within Z-space of both electromagnetic quanta of the electron in motion "B" and "C", to be correlated with the resonance mechanics symbolically represented in **Figure 2.8** and **Table 2.1**.

B – Symbolic representation of the spherical volume of the oscillating magnetic energy of the electron in Z-space. Ref. **Figure 2.5-c** as applied to the electron inner oscillating structure and Equation (2.53). This volume corresponds to its magnetic energy varying from zero presence to maximum presence calculated with Equation (2.22), and to half its invariant rest mass as determined in Reference ([30], [8] Chapter 4).

C – Symbolic representation of the spherical volume of the oscillating magnetic energy of the electron carrier-photon in Z-space. Ref. **Figure 2.5-c** as applied to the carrier-photon inner oscillating structure and Equation (2.55). This volume corresponds to its magnetic energy varying from zero presence to a maximum presence calculated with Equation (2.23), and to the velocity related electron magnetic mass increment Δm_m as calculated with Equation (2.10). This volume also corresponds to the energy contained in the volume defined by the Schrödinger wave function.

D - Central resonance anchoring point $\otimes$ of the magnetic energy of the electron within resonance volume "A", that is, its trispatial junction point, where the origin of the trispatial complex is located for the electron energy quantum (**Figure 2.4**).

E - Central resonance anchoring point $\otimes$ of the magnetic energy of the electron carrier-photon within resonance volume "A", that is, its trispatial junction point, where the origin of the trispatial complex is located for the carrier-photon energy quantum (**Figure 2.4**).

Note that more realistically, the combined magnetic volume of both magnetic quanta should amount to a single spheroid whose dimensions would vary as a function of the constantly varying sum of the magnetic energies present in Z-space at any given instant due to their constant alternance between maximum presence and zero presence at different frequencies, and within which both anchoring points "D" and "E" would physically remain at some varying distance ΔZ from each other by structure as they oscillate toward and away from each other as will be analyzed with **Figure 2.8**. This exploded representation is only meant to help visualize that both quanta oscillate separately at their respective frequencies.

F - Unidirectional orientation along the X-x axis in X-space of the ΔK electron momentum motion energy.

ΔZ -Zitterbewegung distance between both trispatial junctions "D" and "E".

Figure 2.7 should be correlated with **Figure 2.8** that represents the transverse interplay determining the actual zitterbewegung resonance state.

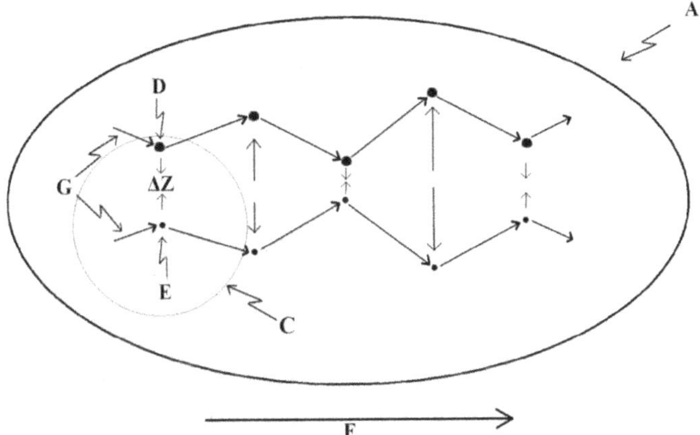

Figure 2.8: Zitterbewegung.

Additional legend for **Figure 2.8**, completing those defined for **Figure 2.7**:

G - The transverse zitterbewegung resonance oscillation resulting from the cyclic relative spin reversal of the magnetic energy spheres "B" and "C" according to their respective frequencies (see **Figure 2.7**) result in an uninterrupted sequence of successive closing in toward each other and moving away from each other of both central resonators anchoring points "D" and "E". The irregularity of the successive cyclic reversed distances is only meant to highlight that both magnetic spheres of **Figure 2.7** cycle from maximum presence to zero presence at different frequencies, resulting in an irregularity of resonance cycles that generate the observed zitterbewegung motion.

As a matter of fact, the relative motion freedom of both tri-spatial junctions $\otimes$ with respect to each other can only be perpendicular to the direction of motion of the system, since the stability by structure of the amount of translational energy of the carrier-photon at any given moment depends uniquely on the Coulomb interaction between the carried electron and other charged particles. This constraint thus prevents any longitudinal deceleration or acceleration relative to each other from being involved in their motion.

The only remaining possible direction of motion available for the two trispatial junctions with respect to each other is thus transverse with respect to the direction of motion of the system, which implies that at any given moment, both trispatial junctions will be at distance ΔZ (zitterbewegung distance) from each other (see **Figure 2.7**), computable as a function of the state of the electromagnetic harmonic oscillations parameters of both quanta at this moment.

Haven't we just identified here the cause of the *zitterbewegung motion* that Schrödinger was discussing in his analysis of the Dirac wave equation [69] and that he concluded amounted to an irregular circular fluctuating motion of the electron, which is superimposed to its translational motion? The difference with Schrödinger's description is that while QM asserts that the magnetic spin moment is caused by the zitterbewegung motion (observed, but unexplained), the trispatial geometry approach predicts and mechanically explains that it would be the zitterbewegung motion which would be due to the forced interaction between the pre-existing cycling magnetic energy of the electron and the pre-existing cycling magnetic energy of its carrier-photon, due to their frequencies differences.

So, on top of revealing that the actual resonance volume *visited* by the two interacting oscillating electromagnetic quanta of the electron in motion will vary with the varying frequency of the increasing or decreasing energy of the carrier-photon due to the varying proximity of the carried electron with other charged particles in its environment, this analysis reveals that when the energy of the carrier-photon becomes exactly equal to that of the invariant rest mass of the electron, that is 0.511 MeV, the amplitude ΔZ of the zitterbewegung oscillation should fall to zero due to perfectly coinciding anti-parallel magnetic spin alignment while the resonance volume synchronizes in simple harmonic oscillation, which could be verified experimentally.

2.19. The wave function and the resonance state of the moving electron

This brings us to put in perspective the resonance volume defined as the moving electron's fixed amount of harmonically oscillating energy interacts with the varying amount of its harmonically oscillating carrier-photon energy as the electron moves in space, with respect to the traditional form of the wave function being used to represent it.

As mentioned at the beginning of this paper, the wave function was initially introduced to represent the resonance volume that de Broglie had concluded that the electron had to be captive into when stabilized in the hydrogen ground state [58]. The method was then extended to represent electrons and electromagnetic photons in free motion.

As it currently stands, the Schrödinger wave function involves the complex harmonic oscillation of an unclearly defined single resonator mathematically combining a real and an imaginary part, whereas from the trispatial geometry perspective, we observe that the electron in motion involves two very clearly defined electromagnetic resonators in separate simple harmonic oscillation.

Even if as it stands, although the Schrödinger wave function allows accounting for the complete complement of momentum energy ΔK of the electron in motion or captive in atomic orbitals, it is unable to account for the electron zitterbewegung motion as stemming from the electromagnetic properties of the electron and of its carrying energy.

Indeed, its classical mechanics origin incompletely related to electromagnetism does not allow reverse engineering any of the electromagnetic characteristics of the resonating electron from this wave function characteristics, which is the disconnect that Feynman observed in 1964 that prevented Quantum Mechanics from being completely synchronized with electromagnetism [26]:

> *"There are difficulties associated with the ideas of Maxwell's theory which are not solved by and not directly associated with quantum mechanics...when electromagnetism is joined to quantum mechanics, the difficulties remain."*

Let us note here that reverse-engineering the manner in which observed phenomena can be explained is a quite usual method of exploration in scientific circles. Indeed, it possibly is the only effective method, but its minimal condition of success hinges on considering as few arbitrary axiomatic grounding premises as possible, while taking into consideration as many related confirmed experimental observations as can be gathered, and finally no unrelated element.

Faced with this dead end when starting from the wave function characteristics, it appeared logical to attempt reverse-engineering the electromagnetic resonating structure of the electron and of its carrier-photon, not from the characteristics of the wave function as de Broglie attempted to do, but from the well established and well known characteristics of electromagnetic energy, which led to the present solution elaborated from the trispatial geometry perspective.

To get an idea of the challenge that de Broglie was confronted with, let's examine how the nature of the resonator generating a resonance volume well understood in classical mechanics can rather easily be understood by means of reverse-engineering.

Who has not observed with a modicum of curiosity how a guitar string that has just been picked practically *disappears* from sight, particularly in the middle of its length as it vibrates, while transversely *visiting*, so to speak, a very characteristic volume of space, which is its actual *resonance volume* that can be represented by a wave function?

In this case, we obviously know in advance that the resonator is a continuous elastic string tied at both ends, because we can actually see the string when at rest, and even though it seems to disappear when vibrating, we also know that the string still physically exists even if we don't see it as it momentarily oscillates transversely too fast for us to see.

We can also imagine that someone having never seen a guitar nor any other string instrument, but who would be expert in mathematics, being shown the very characteristic wave function describing completely the stationary resonance volume of the string, after carefully observing the symmetrically diminishing toward zero of the amplitude of the resonance volume on either side of its maximum value, may well be able to deduce that this resonance volume could only have been produced by a continuous elastic string anchored in fixed positions at both end, thus discovering and understanding the nature of a resonator that he knew nothing about previously.

But no such luck with the Schrödinger's wave function because, as we saw in the previous section, the electromagnetic resonance anchoring points of his wave function that allow understanding how its resonance mechanics can be established, are not conveniently located outside the resonance volume as in the case of the guitar string, but inside this volume, which provides no clue whatsoever that could help even recognizing their very existence and consequently their relation to electromagnetism. This is why the only possible reverse engineering direction that could reveal the relations between Schrödinger's wave function and electromagnetism was from the confirmed characteristics of electromagnetic energy.

In fact, the identification of the electromagnetic localizations parameters allowed by the trispatial mechanics show that the Schrödinger wave function has been mapping the resonance volume of the ΔK momentum related half-quantum the electron carrier-photon, which means that when the wave function is made to theoretically *collapse*, it is the momentary location in space of the electron carrier-photon trispatial junction "E" which is physically located [18], and its momentary ΔK momentum energy which is revealed. See **Figures 2.7, 2.8** and **2.9**.

The relative position of the electron trispatial junction "D" can then established to lie at distance ΔZ (momentary zitterbewegung distance between both trispatial junctions) from the carrier-photon trispatial junction "E" at the same perpendicular distance from the atomic nucleus when the electron is captive in an atomic orbital (see **Figure 2.9**).

With the help of **Figure 2.7** to establish a mental representation of the related "B"/"C" magnetic interactions, we thus observe that both electromagnetic components are kept together by the sequence of cyclic transverse magnetic attraction/repulsion reversals due to the fact that their separate magnetic energy "B" and "C" are constantly switching between mutual relative parallel and anti-parallel alignments of their magnetic spins at different frequencies ([43], [8] Chapter 2) ([11], See also Chapter 3) ([30], [8] Chapter 4); the spherical magnetic orientation of the electron energy "B" cyclically reversing at the invariant frequency calculated with Equation (2.15), while that of its carrier-photons "C" that varies with the amount of kinetic energy of which it is made, cyclically reverses at the frequency that can be calculated with Equation (2.14).

Each closing in sequence between the "D" and "E" trispatial junctions corresponds to the duration of a phase of magnetic antiparallel alignment of the spins both magnetic spheres "B" and "C", corresponding to the fact that the sum of their energies present in Z-space progressively diminishes toward some momentary minimum presence value, while each moving away sequence corresponds to a phase of magnetic parallel alignment of their spins, corresponding to the fact that the sum of their energies present in Z-space progressively increasing toward some momentary maximum presence value.

Given that both magnetic spheres oscillate at different frequencies, these minima and maxima will vary according to the extended resonance sequence specific to their combination as a function of the variation of the adiabatic energy making up the carrier-photon as it moves in space with reference to varying distances between this electron in motion and surrounding other charged particles, thus completely accounting for the apparently random observed zitterbewegung motion.

2.20. *The resonance states of the electron in atomic orbitals*

As analyzed in References ([43], [8] Chapter 2) ([11], See also Chapter 3), the only way for an electron to be stopped in its motion when moving freely in Nature, is for it to be captured into some least action axial electromagnetic equilibrium states in one of the authorized orbitals in an atom.

During its just analyzed free motion already, both separate electromagnetic quanta making up the electron in motion, that is, that of the invariant amount of energy of its rest mass "D" and that of the energy of its carrier-photon "E", can only be maintained together because the interaction in high frequency cyclic inversion of their magnetic energy "B" and "C", whose attractive presence phases, despite being intermittent and asynchronous as a function of the inverse cube of the distance, is sufficiently strong at so short distances, to insure a cohesion that can only be a state of least action by definition.

But as powerful this interaction can be at so short distances between the magnetic energy spheres "B" and "C", it is dwarfed out of all proportions with respect to the power of the interaction between these magnetic energy spheres and the magnetic energy spheres "N" of the carrier-photons of the up and down quarks making up the proton constituting the nucleus of a hydrogen atom (See **Figures 2.9** and **2.10**).

So powerful in fact, that even at the *relatively astronomical* approximate distance of 5.29E-11 m from the proton, the complex resultant of their combined cyclic repulsive parallel magnetic interaction is sufficient to literally stop the electron in its tracks as it is in the final leg of its acceleration motion toward the proton, the latter due to the Coulomb force attraction between its negative charge and the positive combined charges of the three quarks, and that the complex resultant of their combined cyclic attractive antiparallel magnetic interaction is sufficient to keep it from escaping and keep it captive in a stabilized least action axial electromagnetic equilibrium state. See Section 1.28.

The parameters of Tables **2.2** and **2.3**, and Equations (2.61) to (2.64) indeed allowed calculating in References ([43], [8] Chapter 2) ([11], See also Chapter 3) ([32], [8] Chapter 14) ([49], [8] Chapter 9) that the magnetic energy component "N" of the energy of each of the quarks' carrier-photons is more than 600 times more powerful than that of the invariant magnetic energy of the rest mass of the electron "B", which mutually build up their combined strength to about 2000 times that of the electron and its carrier photon.

During the actual stopping process, the forward moving ΔK momentum related energy half-quantum "F" of the electron carrier-photon had no option, due to its forward inertia, but to escape as a well known Bremsstrahlung electromagnetic photon, whose energy amounts is 13.6 eV in the case of the establishment of the electron in the hydrogen ground state orbital "H". See Section 1.28 for the detailed emission mechanics of this photon in the trispatial geometry.

As this momentum energy escapes, the exact same amount of replacement ΔK momentum energy "F" is simultaneously adiabatically re-induced by the Coulomb force as described in Reference ([43], [8] Chapter 2), because it is well verified that the Coulomb interaction between charges forbids that an amount different from 27.2 eV be induced as a carrier-photon in unit charges separated by this distance of 5.29E-11 m.

This new half-quantum of ΔK momentum energy "F" now directly and unswervingly oriented by structure toward the proton will continue applying a continuous *pressure* to keep the electron negative charge moving toward the oppositely signed resultant of the charges of the nuclear subcomponents, even if its forward motion is impeded by the magnetic counter pressure existing between its magnetic energy "B" and that of the inner proton carrier-photons "N".

And it is the *pressure/counter-pressure interplay* between the electron carrier-photon momentum energy ΔK "F" and the complex interaction between the oscillating magnetic spheres "B" and "N" involved that determines the resonance volume described by the Schrödinger wave function, as we will see.

It must be said that the capture of an electron by a proton to form a hydrogen atom is possibly the best-understood process involving elementary particles stabilized into least action axial electromagnetic equilibrium. It has however been studied and understood over the course of the past century only by means of the two traditional distinctly different filtering paradigms that cannot be directly reconciled, that is, that of classical/relativistic mechanics physics and that of quantum mechanics physics.

From the classical/relativistic mechanics paradigm, inherited from Newton's mechanics, stabilization of the electron at the calculated distance of 5.291772083E-11 m can be related only to the idea that the electron would be a localized mass without internal structure orbiting the proton at this distance at the velocity corresponding to its ΔK momentum energy, a velocity that can be calculated either from the classical or the relativistic viewpoint depending on whether or not the gamma factor is taken account of in its calculation.

From this perspective, it is not conceivable that the electron could conserve its ΔK momentum kinetic energy as calculated with Equation (2.11) if it were to slow down and become immobile at this axial distance from the proton, because the very existence of kinetic energy, from the classical/relativistic mechanics perspective, depends on the velocity of a massive body ([11], See also Chapter 3). From this perspective, if a massive body were to slow down in this manner, its kinetic energy is deemed to convert to an equivalent amount of *potential* energy, which would be tantamount to depriving the electron of any possibility of remaining *in orbit*, and it is considered that this would lead to the electron theoretically *falling* onto the proton.

But of course, since we know with certainty that this never happens in physical reality, from countless experiments carried out over the course of the past century, we also know that this conclusion, drawn with reference to macroscopic massive bodies before the existence of electric charges and of the Coulomb force were discovered, is somehow at least partly misleading when applied to the behavior of electric charges, even if it appears satisfactory when applied to massive bodies at our macroscopic level.

From the quantum mechanics perspective on the other hand, inherited from the establishment of the Schrödinger wave function and Heisenberg's statistical distribution in the 1920's, the electron stabilized in the hydrogen ground state is seen with 100% probability as being present within a clearly defined resonance volume of space about the proton, within which the energy of the electron, without any internal structure just like in classical/relativistic mechanics, is estimated to statistically be more concentrated (or more often present) about this mean distance of 5.29E-11 m from the proton, a volume within which the electron cannot be seen as moving on a clear trajectory contrary to classical/relativistic mechanics, even though it is clearly established that it can be axially located anywhere within this volume when any theoretical wave function collapse is calculated, and that its most probable location tends to coincide with the classical Bohr orbit representation, which is expressed as a probability of increased density of the energy of the electron within the volume described by Heisenberg's statistical method.

Its total energy is defined in more general terms with the Hamiltonian inherited from the classical mechanics paradigm as combining in a single conservative concept the sum of the kinetic energy and the potential energy that accounts for is momentum in classical/relativistic mechanics, intriguingly still internally grounded on the same $p=mv$ Newtonian conservative momentum concept ($p=\gamma mv$ from the relativistic perspective), that causes the ΔK amount of related kinetic energy to still depend on velocity, even if no velocity can be associated with the spread out energy of the electron as currently represented by the wave function resonance volume.

Although both traditional paradigms take into account the ΔK momentum kinetic energy amount from Equation (2.11), neither of them takes into account the energy corresponding to mass increment Δm_m from Equation (2.2), despite its proven existence experimentally confirmed by the Kaufmann experiments [36] as measured by means of transverse interaction, and consequently, this is why neither paradigm assigns any function to the magnetic fields of charged particles nor to their magnetic mass increments in submicroscopic interaction.

This pinpoints exactly where the disconnect resides between classical/relativistic mechanics and quantum mechanics on one hand, and electromagnetic mechanics on the other hand, and reveals the importance of the adiabatic nature of energy induction ([43], [8] Chapter 2) by means of the Coulomb interaction as shown with **Figure 2.1** and Equation (2.20), that combines as Equation (2.13) the total amount of energy adiabatically induced in charged particles as calculated with Equation (2.11) for the translational momentum component, and with Equation (2.2) for the magnetic mass increment.

The critical disconnect resides precisely in the fact that the ΔK momentum kinetic energy half of the total energy quantum induced is adiabatically induced by the Coulomb interaction (Equation (2.12) in such a way that it can only remain physically present and vectorially active in the direction of the proton, even if it is experimentally proven not to be able to cause the electron to move forward toward the proton, nor along the trajectory mandated by classical mechanics since its vectorial orientation is immutably set by structure perpendicularly to this classical trajectory.

This brings to attention the fact that the relativistic momentum energy ΔK of Equation (2.6) and the Δm_m relativistic mass increment from Equation (2.2) as combined in Equation (2.13), that completely account for the relativistic velocity and relativistic mass increase confirmed by the Kaufmann experiment [36], remain entirely adiabatically induced even when the related relativistic velocity is prevented by *something* from being expressed when the electron is stabilized in the hydrogen atom ground state.

This in turn leads to the conclusion that terms such as *electromagnetic momentum* and *magnetic mass increment* would be more appropriate than the current terms *relativistic momentum* and *relativistic mass increments* to describe these adiabatically induced energy half-quanta since it can be demonstrated that electromagnetic energy is adiabatically induced strictly as a function of the distance between charges, according to the induction growth curve dependent on the gamma factor and the Coulomb force ([11], See also Chapter 3) [36] ([42], [8] Chapter 5), and that contrary to the very foundation of all traditional theories about energy and matter exclusively elaborated from macroscopic level experiments, according to which kinetic energy can exist only if translational motion is possible, kinetic energy is found, from all experiments involving submicroscopic charged elementary particles, to be a *physically existing substance* whose existence does not depend on velocity as currently axiomatically assumed, but that it is velocity that depends on the prior existence of kinetic energy, a velocity that can be expressed only if charged particles' translational motion is not impeded by local magnetic translational counter-pressure ([43], [8] Chapter 2).

The ultimate question then turns out to be: How could this *something* operate to so effectively and so systematically hinder the natural motion of the ΔK

momentum kinetic energy of the electron in such a way that it makes it impossible for it to go crashing onto the proton in accordance with its natural vectorial orientation?

Neither classical/relativistic mechanics nor quantum mechanics offer any mechanical clue to resolve this issue. But the trispatial geometry allows observing that this hindrance can only be provided by a predominantly repulsive magnetic interaction, that is, a magnetic counter-pressure, resulting from of the constant parallel/anti-parallel magnetic spins orientation switching interaction between the magnetic energy "B" of the electron invariant rest mass and those of the 3 quarks carrier-photons "N" of the proton ([43], [8] Chapter 2) ([11], See also Chapter 3) ([49], [8] Chapter 9), as symbolically represented with **Figures 2.9** and **2.10** that we will now analyze.

It must clearly understood that it is the spherical increase/decrease motion of the physical presence of the actual *magnetic energy substance* of the electron and of the 3 quarks carrier-photons that must be visualized during this analysis, and not that of their mathematical E and B fields representations of Maxwell's equations, as we would be intuitively tempted to do.

To really understand the axial resonance trajectory that the electron is forced to move into, that determines the volume defined by the Schrödinger wave function, the relative powers of the oscillating magnetic spheres involved must be put in perspective.

In this process, the magnetic half-quantum Δm_m of the electron carrier-photon "C" will be ignored to simplify the current analysis, because it is infinitesimal in the definition of the ground state resonance volume when compared to the role played by the magnetic mass "B" of the electron as revealed by its value calculated with Equation (2.27) and the following ratio established with the magnetic mass of the electron stabilized in the hydrogen ground state, and is significant only with respect to the transverse zitterbewegung motion of the electron previously analyzed:

$$\frac{\Delta m_m}{m_e/2} = \frac{2.4253377726\text{E}-35}{4.55469094\text{E}-31} = \frac{1}{1.8779615277\text{E}4} \tag{2.65}$$

The ΔK momentum half-quantum "F" of the electron carrier-photon does have a role to play however, because each time that the magnetic sphere "B" of the electron rest mass reduces to zero presence within Z-space, all magnetic counter-pressure disappears by structure between the electron and the magnetic energy making up the magnetic spheres "N" centered on the proton, which causes the ΔK momentum energy "F" of the electron to be free again to propel the electron toward the proton, until the magnetic energy substance "B" of the electron rest mass begins to increase again in Z-space as the following cycle of its frequency initiates.

On the side of the proton, it is the magnetic masses "O" of the up and down quarks that will be ignored, because contrary to the insignificance of the magnetic Δm_m half-quantum "C" of the electron carrier-photon with respect to the magnetic mass of the electron as shown with Equation (2.65), it is the magnetic masses "O" of the up and down quarks that are insignificant when compared to the immensely larger values of the Δm_m magnetic half-quanta of their carrier-photons. Indeed, as calculated in Reference ([32], [8] Chapter 14) each quark carrier-photon "N" would have a mean total energy of about 310.457837 MeV:

$$\text{Quark carrier-photon energy} = \Delta K + \Delta m_m = 4.974082389\text{E}-11\,\text{j} \tag{2.66}$$

which sets their frequency and wavelength to the following values:

$$v = \frac{E}{h} = 7.506837869\text{E}22\,\text{Hz} \qquad \lambda = \frac{c}{v} = 3.993591752\text{E}-15\,\text{m} \tag{2.67}$$

and even without taking into account the magnetic drift caused by the so close mutual proximity of the 6 inner electron electromagnetic quanta of the proton (Ref: Equation (2.62) and References ([32], [8] Chapter 14) ([49], [8] Chapter 9), that considerably increases their magnetic energy, to simplify this analysis, each of their Δm_m magnetic half-quanta will minimally have the following value:

$$\Delta m_m = \frac{E/2}{c^2} = 2.767206524\text{E}-20\,\text{kg} \tag{2.68}$$

In relation with the up quark mass available in **Table 2.2**, the following ratio can be established:

$$\frac{\Delta m_m}{m_U/2} = \frac{2.767206524\text{E}-20}{1.024805462\text{E}-30} = \frac{2.700226166\ \text{E}10}{1} \tag{2.69}$$

and for the down quark:

$$\frac{\Delta m_m}{m_D/2} = \frac{2.767206524\text{E}-20}{4.09922189\ \text{E}-30} = \frac{0.6750565347\ \text{E}10}{1} \tag{2.70}$$

So, comparing these last two ratios with the electron magnetic masses ratio calculated with Equation (2.65), not only do we observe that these two ratios are reversed with respect to the relation between the electron magnetic mass and that of its carrier-photon, but we also observe that the quarks carrier-photons are

10 orders of magnitude more energetic than the quarks that they carry, which justifies taking only the magnetic spheres "N" of these 3 carrier-photons into account to summarily explain the electron resonance volume.

Finally, the ratio of the magnetic mass "B" of the electron $m_e/2$ with respect to the minimal Δm_m magnetic mass "N" of even only one of the quarks' carrier-photons, will give a glimpse of how easily and strongly the electron can be set in resonance like a feather in hurricane winds as it is shoved about axially by even the minimal eleven-fold greater magnitude magnetic energy of even only one quark carrier-photon centered on the proton location:

$$\frac{m_e/2}{\Delta m_m} = \frac{4.55469094 \; E\text{-}31}{2.76720652 \; 4E\text{-}20} = \frac{1.64595266 \; E11}{1} \tag{2.71}$$

Facing the single electron oscillating magnetic energy sphere "B", the combined magnetic energy of the inner proton components materialize as two relatively concentric antiparallel energy spheres of unequal spherical volumes (see **Figure 2.10**). The largest is made of the cyclically varying sum of the magnetic energy of two quarks carrier-photons (2 x "N") in permanent mutual parallel spins alignment as it cyclically increases and decreases between zero energy presence and maximum energy presence in Z-space, while the smallest magnetic sphere is made of the magnetic energy "N" of the remaining third carrier-photon, which can only be by structure in antiparallel spin alignment with the first two, and whose energy is in constant oscillation in opposition with the sum of the spherical motion of the magnetic energy of the first two, that is, in increasing energy presence phase while the energy presence of the first two is decreasing, and in decreasing energy presence while the energy of the first two is increasing.

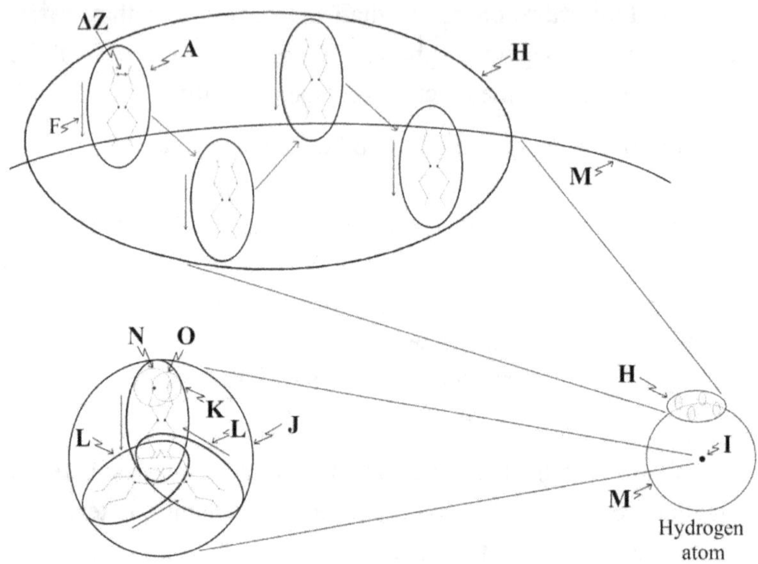

Figure 2.9: The hydrogen atom resonance states.

Additional legends for **Figure 2.9**, completing those defined for **Figures 2.7** and **2.8**.

H - Symbolic representation of the resonance volume, within which the electron point-like behaving trispatial junction and that of its carrier-photon remain captive in the hydrogen ground state orbital.

I - Proton.

J - Symbolic representation of the proton energy resonance volume, resulting from the cyclic spins reversal interaction between all 6 inner components of the proton structure, which are 2 up quarks, 1 down quark and their 3 carrier-photons.

K - Symbolic representation of the zitterbewegung resonance volume within which the point-like behaving down quark trispatial junction and that of its carrier-photon involved in mutual cyclic magnetic spin reversal interaction within Z-space remain captive, as represented with Figures **2.7** and **2.8**, but involving much higher frequencies than the zitterbewegung resonance volume of the electron.

L - Symbolic representation of the zitterbewegung resonance volume within which the point-like behaving up quark trispatial junction and that of its carrier-photon remain captive according to the same mechanics described for the down quark with previous representation K.

M -Hydrogen ground state mean orbital distance between the electron and the proton, corresponding to the theoretical Bohr radius at which the electron momentum energy is set at precisely ΔK, outside of which distance this momentum energy diminishes to $\Delta K -\Delta(\Delta K)$ when the electron is pushed further away, and increases to $\Delta K +\Delta(\Delta K)$ when it is pulled closer to the proton, during its cyclic axial resonance motions sequences.

N –Symbolic representation of the spherical volume of the oscillating magnetic energy of a quark's carrier-photon in Z-space. Ref. **Figure 2.5-c** as applied to the carrier-photon inner oscillating structure and Equation (2.55). This volume corresponds to its magnetic field varying from zero presence to a maximum presence that can be calculated with Equation (2.23), using the carrier-photon wavelength obtained with Equation (2.67), and to the velocity related quark magnetic mass increment Δm_m obtained with Equation (2.68).

O –Symbolic maximum extent of the spherical volume of the oscillating magnetic energy of one quark up or down in Z-space. Ref. **Figure 2.5-c** as applied to the quark inner oscillating structure and Equations (2.63) and (2.64).

Since all three quarks carrier-photons have the same frequency, they remain permanently synchronized in one of the two possible configurations, which are "U $\parallel$ U $\neq$ D" or "U $\parallel$ D $\neq$ U". See **Figure 2.10**.

The outcome is that irrespective of which increasing or decreasing presence phase of its magnetic oscillation "B" the electron may be in, either the larger or the smaller magnetic sphere centered on the proton location will be in parallel spin alignment with it and will repel it, which is what permanently interdicts that the electron could naturally reach the proton, unless having been accidentally or artificially induced with sufficient energy from outside sources, as is routinely done in high energy accelerators.

In the symbolic representation of **Figure 2.9**, the electron zitterbewegung resonance volume "A" is oriented as if it was going to move toward the proton, to reflect the fact that the 13.6 eV half-quantum of momentum energy ΔK re-induced in the electron carrier-photon as the electron was captured remains permanently oriented toward the proton, even if its forward motion is constantly inhibited by the fact that whatever increasing or decreasing presence phase its magnetic oscillating sphere may be in (Characteristic "B" in **Figures 2.7** and **2.10**), the latter will be repelled since either one or the other of the two mutually antiparallel magnetic spheres centered on the proton location (See **Figure 2.10**) will always be in repulsive parallel spin alignment with respect to the electron magnetic energy sphere "B".

This constant relative parallel repulsive spin orientation of the electron magnetic sphere with respect to at least one of the two proton concentric magnetic spheres is not however what explains the ground state orbital axial resonance state of the electron defined by the Schrödinger wave function. We will get into this in a moment, but let's first analyze the proton electromagnetic structure.

There may appear to be a disconnect between the idea that the energy of the magnetic oscillating spheres of the quarks carrier-photons could reach as far in space as the ground state orbital located at 5.29E-11 m from the proton, and with sufficient strength at that to establish a least action electromagnetic equilibrium state keeping the electron captive at this relatively great distance from the relatively minuscule volume of radius 1.2E-15 m within which we know that the 6 inner electromagnetic quanta making up the proton are captive into.

This is more easily put in perspective when considering that the magnetic field of the Sun reaches out as far as the outer limits of the Solar System, presumably as far as the Oort cloud, even if the matter of which the Sun is made is contained within a sphere whose radius is well shorter than the radius of Mercury's orbit, its innermost planet. Indeed, the huge magnetic field of the Sun can only be made of the sum of the individual magnetic fields of the innumerable elementary particles and carrier-photons making up the matter of which the Sun is made. The same conclusion can obviously be drawn for all existing celestial bodies, as put in perspective in Reference ([45], [8] Chapter 16).

So there is no disconnect between this conclusion drawn at the submicroscopic level and what can be observed even at the astronomic scale, because if a hydrogen atom was theoretically upsized sufficiently for its proton diameter to reach in dimension that of the Sun, as often reminded of during these analyses, then the electron would stabilize as far as Neptune's orbit, which relatively speaking, would give the magnetic field of the proton the same order of magnitude as that of the Sun.

Let us also recall that in the trispatial geometry, it is not within normal X-space that this magnetic energy expands and contracts, but within magnetostatic Z-space, and that only the ΔK momentum energy half-quanta and the point-like trispatial junctions $\otimes$ of each elementary particle, free moving photon and carrier-photon that actually really *live*, so to speak, within normal X-space, that is, the only two aspects of elementary particle's electromagnetic energy that are physically detectable by means of frontal longitudinal collisions, which are the total sum of the electromagnetic energy that resides in the other two orthogonal spaces Y and Z, and whose physical presence we can detect only through these trispatial junctions $\otimes$ that behave point-like in X-space, and their ΔK translational momentum energy; and the only aspect of electromagnetic energy that can be detected by transverse collision or interaction, which is only the electromagnetic energy that resides in the other two orthogonal spaces and that we detect through these trispatial junctions $\otimes$ always behaving point-like in X-space, that is, the energy of the rest masses m of the electron, the positron, the up quark and the down quark, and the energy of the magnetic mass increments Δm_m of the carrier-photons, and finally the electromagnetic energy half-quanta Δm_m of freely moving photons.

Indeed, it is not the magnetic fields of the proton 6 inner components that are captive within its physically measured volume, but the 6 point-like behaving trispatial junctions $\otimes$ that are the individual anchoring locations of this magnetic energy within normal X-space, and through which their electromagnetic energy cyclically oscillates, that are captive by pairs in zitterbewegung transverse resonance states, and also collectively in the common least action resonance volume resulting from their mutual trispatial electromagnetic interaction resulting in the establishment of the stable proton structure.

The neutron, which is not illustrated in this document, has an inner electromagnetic structure involving the same up and down quarks and their carrier-photons of slightly higher energy, with the difference that instead of involving 2 up quarks and 1 down quark (uud), it involves 2 down quarks and 1 up quark (udd). The details of the trispatial structures of both nucleons are available in Reference ([32], [8] Chapter 14).

With regard to the ground state orbital resonance volume, **Figure 2.10** puts in perspective the fact that this resonance volume is due to the hugely higher oscillating frequency of the quarks carrying-photons energy with respect to the much slower oscillating frequency of the electron rest mass magnetic energy.

Relating these frequencies of the electron from Equation (2.15) and of a quark carrier-photon from Equation (2.67) allows determining that minimally, the magnetic polarity reversal of each quark carrier-photon occurs in excess of 600 times during each occurrence of magnetic polarity reversal of the electron magnetic energy, that is, during each magnetic presence cycle of the electron magnetic energy in Z-space:

$$\frac{v_{quark\,carrier\text{-}photon}}{v_{electron}} = \frac{7.506837869\,E22}{1.235589976E20} = \frac{607.5508878}{1} \tag{2.72}$$

The constant interplay due to the frequencies difference of the various magnetic spheres involving the inverse cube interaction law with distance, that opposes the ΔK unidirectional momentum energy "F" that constantly tends to propel the electron toward the proton, to an uninterrupted sequence of magnetic attraction-repulsion phases, can then only result in the establishment of the stable axial resonance state that de Broglie identified [58].

In **Figure 2.10**, the central sequence "B" symbolically represents an arbitrary sample of 6 occurrences of the intensity variation of the spherical presence of the electron magnetic energy as a function of its frequency. In a simplified manner, each of these 6 occurrences is confronted in the lower sequence by the more than 600 occurrences of the intensity variation of the spherical presence of the magnetic energy of the 3 carrier-photons of the up or down quarks of the proton as a function of their own frequencies.

The least action orbital equilibrium state is consequently established by the fact that the ΔK momentum energy "F" half-quantum of the electron carrier-photon, is alternately hindered in its forward motion, when the magnetic interaction function of the inverse cube law becomes repulsive – parallel magnetic spin alignment between the magnetic energy spheres of the electron and one of the proton magnetic energy spheres – and is then freed from this counter-pressure while the magnetic interaction becomes attractive – antiparallel magnetic spin alignment between the magnetic spheres involved.

As represented with **Figure 2.10**, during each of the 600 magnetic cycles of a quark's carrier-photon "N", the electron magnetic sphere "B" will be axially repelled away from the proton by distance Δd during half of the carrier-photon "N" magnetic presence cycle, during which their spin alignment is parallel –

thus repulsive, and since the electron will be farther away from the proton as the relation becomes antiparallel for the same duration, there will be a physical impossibility for it to be axially brought back all the way to distance $-\Delta d$, given that the inverse cube force will be weaker at this farther location from the proton at the beginning of the antiparallel phase.

Therefore, and by structure, given the more weakly acting inverse cube attraction at the beginning of attractive phase, the electron can be axially brought back only to distance $-(\Delta d-\Delta(\Delta d))$, which will cause it to progressively move away from the proton at each "B"/"N" relative magnetic spins polarity reversal sequence until its own magnetic energy presence "B" falls to zero, moment during which only the electron carrier-photon ΔK half-quantum momentum energy will be active, now causing the electron to freely move as close to the proton as the Coulomb force inverse square law will bring it, until its next magnetic presence cycle "B" initiates and that the whole predominantly repulsive magnetic sequence "B"/"N" is initiated again, as represented with **Figure 2.10**.

Of course the actual resonance state of the electron in the least action orbital of the hydrogen atom or in any other atom will be much more complex than hinted at with this limited example, which is only meant to describe the fundamental mechanics of the magnetic interaction between the electron magnetic energy "B" and ΔK momentum energy on one hand, and the magnetic energy of the quarks carrier-photons of the proton, on the other hand. Obviously, the exact resonance volume within which each elementary electromagnetic massive particle in the hydrogen atom will be circumscribed, which are one electron, one down quark and two up quarks, can eventually be determined only by a careful study of all electromagnetic interactions between them and their carrier-photons.

Given that the mean equilibrium distance that this process forces the electron in motion to stabilize at in the hydrogen atom coincides with the densest area of probability distribution of Heisenberg's statistical method, it would seem that the axial trajectory of the electron about this mean distance within the volume that the electron can thus visit as a function of its varying relativistic mass and related inertia at any given instant, it should directly correspond with Heisenberg's probability distribution of all of the possible instantaneous locations that the electron can be stochastically calculated to be localized at when repeatedly theoretically collapsing the wave function in its current form

[36] [67], and whose quantized axial beat can no doubt be related to the regularities of the fine structure of the hydrogen spectrum, that Sommerfeld first associated to a hypothetical elliptical orbit that the electron would follow, in his attempt to explain the fine splitting of the main spectral lines ([18], p.114).

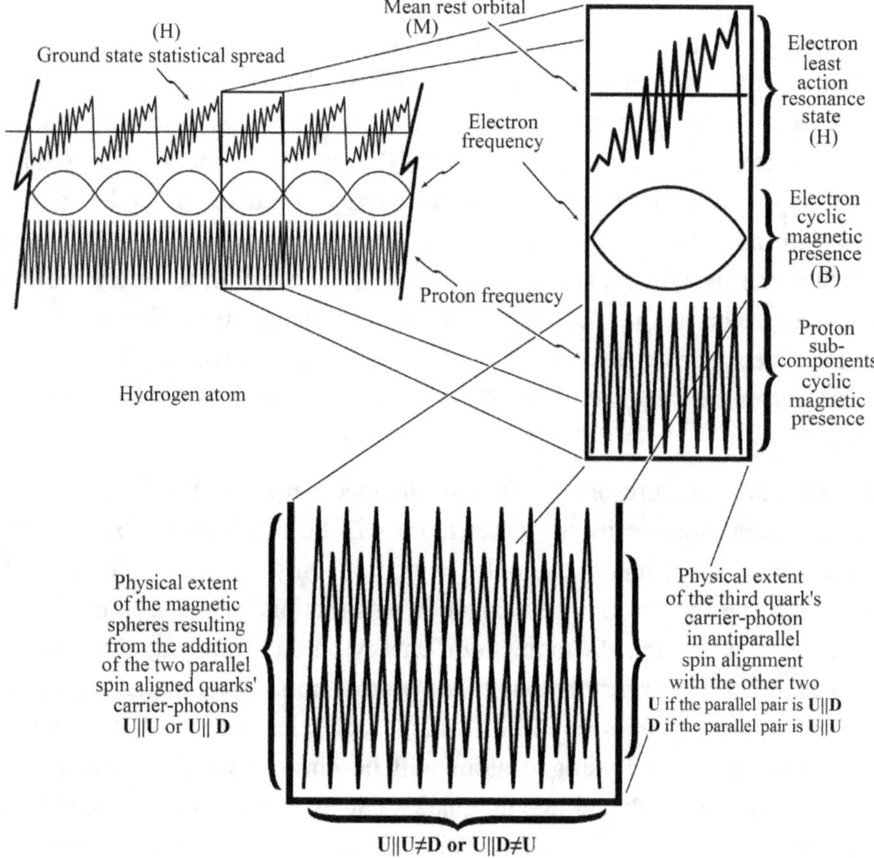

Figure 2.10: Establishment of the least action resonance state of the electron in the hydrogen atom.

So the very limited resonance volume in X-space within which the ΔK momentum and the trispatial junctions $\otimes$ of a moving electron will be localized within can be represented as:

$$\int_{-d}^{+d} |\psi|^2 \, dxdydz = 1 \tag{2.73}$$

while the theoretical resonance volume in Z-space within which the dynamically oscillating magnetic energy sphere of the same moving electron can be represented to have resonance repercussions, due to mutual contacts between it and all other existing dynamically oscillating magnetic energy spheres in existence in the universe, would be:

$$\int_{-\infty}^{+\infty} |\psi|^2\, dxdydz = 1 \tag{2.74}$$

It seems also entirely reasonable to conclude that the elementary charged up and down quarks making up the scatterable inner structure of protons and neutrons and their carrier-photons, which are known to be the only existing elementary electromagnetic subcomponents of all atomic nuclei, as analyzed in Reference [36] ([32], [8] Chapter 14), should be subject to similar resonance states within their own local least action electromagnetic equilibrium states, that could then also potentially be described by the various methods of quantum mechanics in a manner more satisfactory than Quantum Chromodynamics (QCD) has achieved.

2.20.1. Interacting atomic and molecular resonance volumes within magnetostatic Z-space

It is assumed that it is the electronic orbitals that define the real spherical volume occupied by atoms in normal X-space, but from the viewpoint of the magnetostatic Z-space, it would rather seem that it is the intense elastic magnetic field in resonance of the atomic nuclei that really defines the atomic volumes in this space, as can be gathered from the stabilization mechanics of the electron in its rest orbital in the hydrogen atom just analyzed, while the elastic magnetic double spheres of the electron enter into a stable resonance state with respect to the six elastic magnetic spheres of the proton, in the allowed orbitals located at specific mean distances from the centre of the nuclear magnetic fields, in the same way that the planets in the solar system are stabilized in the various stable orbits allowed within the magnetic field of the Sun whose size encompasses the entire solar system.

According to this perspective, the entire universe would thus be populated, from the subatomic to the astronomical level, by innumerable magnetic spheres of elementary particles oscillating electromagnetically, interacting elastically in antiparallel magnetic alignment to capture each other, or combining in forced parallel magnetic alignment to merge into macroscopic static magnetic fields, whose interactions establish the full hierarchy of stable axial stationary equilibrium resonance states that can be observed, in conjunction with the counterpressure exerted by the momentum energy of these particles' carrier-photons, which is always oriented vectorially so as to counteract the default mutual repulsion resulting from the parallel spin orientations due to their different oscillation frequencies. See Sections 3.19 to 3.21.

For example, the electrons of two hydrogen atoms naturally combine in a state of antiparallel magnetic spin, since this is their least action state to form, thanks to this very strong covalent bond, a H_2 hydrogen molecule. The result obtained is the association of the two composite magnetic spheres of the two protons repelling each other on either side of the covalent electronic bound due to both their same sign mutually repelling electric charges, and their presumably dominating default permanent mutually repelling parallel spin states.

The resonance frequencies of a hydrogen gas volume — composed of numerous H_2 molecules in mutual interactions, or of helium gas – composed of helium atoms in mutual interaction, in which the nuclear magnetic fields are completely exposed to interact with other atoms or molecules, remains to be analyzed according to this perspective.

The same also for all molecules comprising hydrogen atoms, whose proton is also exposed to other atoms or molecules, contrary to higher atomic weight atomic atoms and molecules whose surrounding electronic escorts is predominantly what interacts with other atoms or molecules, like for example, the water molecule, that combines both two exposed protons and an exposed predominantly electronic sphere.

2.21. Conclusion

Although not providing the progressive mechanical explanation of the transitions between stationary states that de Broglie and Schrödinger were working to resolve and that Sections 1.28 and 1.29 analyze further, article [10] reproduced in this chapter proposes an electromagnetic resonance mechanics of stabilization of the electron in the hydrogen atom ground state that, if confirmed, could possibly allow it. However, Chapter 1 of this book, that reproduces article

[9], does indeed propose a mechanics of electromagnetic photon emission and absorption from the trispatial perspective that can help clarify these transitions.

Unexpectedly, this mechanics also brings to light the possibility of using the various methods of quantum mechanics to establish wave functions to describe the resonance states of the elementary particles making up the inner scatterable structure of protons and neutrons.

This mechanics of establishment of the resonance volume representable by a wave function is grounded on the identification of the anchoring points $\otimes$ inside this volume of both quanta of electromagnetic energy constituting an elementary electromagnetic particle, that is, the trispatial junctions through which the quanta of kinetic energy involved oscillate harmonically to establish this volume.

The trispatial geometry reveals that each stable elementary electromagnetic particle involves in fact a pair of separate electromagnetic quanta, that is, a massive stable elementary electromagnetic quantum which is intrinsically translationally inert in X-space – electron, positron, up quark and down quark – possessing a measurable electric charge and a measurable magnetic field, accompanied by a carrier-photon, possessing a pair of electric charges whose opposite signs mutually cancel each other and a measurable magnetic field, and which contributes the momentum ΔK of the inert quantum that it accompanies in space, as well as its electromagnetic mass increment Δm_m.

This geometry reveals furthermore that the spin of elementary particles is a property of relative alignment of magnetic polarity between the electromagnetic particles and not an intrinsic property of angular moment of these particles, and that the half-quantum of magnetic energy of any electromagnetic quantum oscillates between a state of maximum presence and a state of zero presence in Z-space at the frequency of its energy.

Finally, the trispatial geometry reveals that it is the differences of frequencies of oscillation of the half-quanta of magnetic energy of elementary quanta that explain the stability of all least action orbitals in atoms, as well as all of their resonance states.

3. Gravitation, Quantum Mechanics and the Least Action Electromagnetic Equilibrium States

3.1. Introduction

The century old challenge of fundamental physics has been to reconcile quantum mechanics (QM) that deals with submicroscopic interactions between elementary particles from the quantization perspective, with relativistic mechanics that deals with gravitation at the macroscopic level from the infinitesimally progressive perspective, mainly embodied by the theory of general relativity (GR). The ease with which infinitesimally progressive sequences of motion can be mathematically represented by means of an indefinite number of instantaneous momentary excited states of a postulated underlying neutral energy quantum vacuum field, which is the foundation of quantum field theory (QFT), has naturally privileged this quantization perspective in all past attempts at reconciling QM with gravitation. But, given that all scatterable elementary particles identifiable within atomic structures have an electrical charge, and are thus electromagnetic in nature, this article explores the possibility of reconciling quantum mechanics with relativistic mechanics from the electromagnetic perspective, by means of reconciling the wave function with the least action electromagnetic resonance states into which elementary charged particles become captive within atomic and nuclear structures, and ultimately, with gravitation.

This chapter reproduces article [11] that was the final installment of a series of papers published in 2000, 2007, 2013, 2016 and 2017, describing the various aspects of an entirely new paradigm of fundamental physics, all of which are synthesized in a monograph published in 2017 by Scholars' Press [8] upon invitation by the editors.

The necessity of being able to refer to clear and concise explanations of each specific aspect of this new perspective resulted in the progressive publication of numerous separate papers, all of which having individually been peer-reviewed and accepted for publication, each of which relating in a self-contained manner a specific aspect of the new paradigm to the traditional paradigm.

The reader will certainly understand that the content of a 600 pages monograph synthesizing and completing the texts of about 20 separate papers certainly would be easier to explore via a simplified general overview, which the present paper is meant to provide.

As mentioned in the Foreword, Article [11] being reproduced as the present chapter, was selected to be republished as Reference [12], as a chapter in the eBook titled "*Prime Archives in Space Research*" by the *Vide Leaf Prime Archives* group, whose aim is to promote scholarly research to the world by regrouping articles considered of value, in view of serving young researchers in the relevant fields to apply these findings to their research practices.

3.2. Maxwell Equations and Mutual Induction of Electric and Magnetic Fields

The new paradigm is entirely grounded on an aspect of electromagnetic theory that has become obscured over time, due to the generalizing perspective afforded by the use of the electromagnetic tensor, which represents both electric and magnetic fields as becoming a single entity, that is, *the electromagnetic field*.

The downside of the otherwise usefulness of tensor treatment is that it conceptually obscures the fact that both E and B fields have different properties and represent different aspects of electromagnetic energy; in particular the fact that in physical reality, according to Maxwell's continuous wave theory, both fields can only mutually induce each other as revealed, among other characteristics, by the fact that the Poynting vector provides by structure the average value of the intensity of the product of the intensities of both time varying oscillating fields ([57], p. 989). The Poynting vector indeed reveals that the time varying product of both fields in vacuum can only be constant:

$$S = \frac{EB}{2\mu_0} \tag{3.1}$$

As clearly explained in traditional undergrad textbooks, such as "*University Physics*" by Sears, Zemansky and Young [17], or "*Physics*" by Halliday and Resnick [57], Faraday's law imposes that a time varying magnetic field acts as a source of electric field. This process is put in practice in the induction of electromotive force (emf) in inductances and transformers. Similarly, Ampere's law, which is used in charging capacitors and to establish current in conductors, demonstrates that changing electric fields are a source of magnetic fields.

> *"Thus, when either field is changing with time, a field of the other kind is induced in adjacent regions of space. We are thus led naturally to consider the possibility of an electromagnetic disturbance, consisting of time-varying electric and magnetic fields, which can propagate through space from one region to another, even when there is no matter in the intervening region."* ([17], p. 696).

Unfortunately such general and complete undergrad textbooks that were used in giving undergrad students a general knowledge of all aspects of fundamental physics, particularly in preparing them to smoothly transit from classical continuous processes to quantum and relativistic physics, progressively went out of fashion to be replaced by textbooks that barely skim over the classical concepts that were directly extrapolated from the classical equations, which were established by the major discoverers of the past from physical experiments that they actually carried out, and that constitute a pool of mutually converging conclusions about electromagnetic energy whose neglect can only lead to a lessening of our understanding of physical reality.

3.3. Kinetic energy and the Coulomb Law

In traditional undergrad textbooks such as "*Physics*" by Halliday & Resnick, the relation between the momentum related kinetic energy and the interaction between charges due to the Coulomb Force is established in the following manner.

From the electromagnetic Coulomb equation applied to calculating the force between the charges of the electron and the proton in a hydrogen atom, taken as the traditional example, and the force calculated from Newton's second law for motion applied to the electron mass in motion ([57], p. 1192) and ([44], [8] Chapter 7):

$$F = \frac{e^2}{4\pi\,\varepsilon_0 r^2} \quad \text{and} \quad F = ma = m\frac{v^2}{r} \tag{3.2}$$

the following relation is drawn in Halliday & Resnick:

$$\frac{e^2}{4\pi\,\varepsilon_0 r^2} = m\,\frac{v^2}{r} \tag{3.3}$$

which allows calculating the momentum related kinetic energy of the electron equally well with either Newton's kinetic energy Equation or with the Coulomb equation (ref: equation 47-19 in Reference [57]):

$$K = \frac{1}{2}mv^2 = \frac{e^2}{8\pi\,\varepsilon_0 r} \tag{3.4}$$

which is how the kinetic energy sustaining the momentum of charged particles is related to the Coulomb force as a function of the distance separating pairs of charges, because the only variable in the Coulomb equation is r, which is the mean distance separating the electron stabilized in the ground state orbital and the proton in the hydrogen atom, which causes any amount of momentum kinetic energy that a charge may possess to be dependent solely on the distances separating it from other charges. Consequently, the closer charges come to each other, the more momentum related kinetic energy they will be induced with, given that the force is acting as a function of the *inverse* square of the distance.

The issue of potential energy is addressed further on in Section 3.23 dealing with the momentum, the Lagrangian and the Hamiltonian and is completely analyzed in correlation with energy conservation in closed systems in Reference [43].

But there is more! In 1903, Walter Kaufmann was the first experimentalist to relate the gamma factor to energy induction during experiments carried out with electrons moving at relativistic velocities in a bubble chamber by accelerating and deflecting them with a combination of electric and magnetic fields [36], by demonstrating that their transverse mass varied with velocity in conformity with the relativistic equation [74]; experiments that he was carrying out in collaboration with theoreticians Max Abraham [39] and Woldemar Voigt, who is the physicist who initially conceived of the gamma factor [72], better known as the Lorentz factor.

So it seems that the gamma factor was initially experimentally related strictly to energy and mass induction with velocity, an experimental result with which Henri Poincare was in agreement:

"Les calculs d'Abraham et les expériences de Kaufmann ont alors montré que la masse mécanique proprement dite est nulle et que la masse des électrons est d'origine exclusivement

électrodynamique. Voilà qui nous force à changer la définition de la masse; nous ne pouvons plus distinguer la masse mécanique et la masse électrodynamique, parce qu'alors la première s'évanouirait; il n'y a pas d'autre masse que l'inertie électrodynamique; mais dans ce cas, la masse ne peut plus être constante, elle augmente avec la vitesse, et un corps animé d'une vitesse notable n'opposera pas la même inertie aux forces qui tendent à le dévier de sa route, et à celles qui tendent à accélérer ou à retarder sa marche."

Henri Poincare ([75], p. 137).

Translation:

"Abraham's calculation and Kaufmann's experiments then showed that mechanical mass proper is null and that the mass of electrons is exclusively of electrodynamic origin. This forces us to change the definition of mass; we can no more distinguish mechanical mass from electrodynamic mass, because then the first would disappear; there is no other mass than electrodynamic inertia; and in this case, mass can no longer be constant, it increases with velocity; and which is more, it depends on the direction, and a body animated with a notable velocity will not oppose the same inertia to forces tending to deflect its trajectory, and to those who tend to accelerate or slow its progress."

It is a historical fact that these physicists who worked closely with the discoverer of the method never accepted the interpretation that the gamma factor could be axiomatically related later to time dilation and length contraction of bodies with velocity.

This means that not only is the momentum related kinetic energy of the electron induced by the Coulomb force, but the energy that serves to increase the mass of a moving electron is also induced by at least one of the ambient electric and magnetic fields, presumably in context the electric field, given its relation to the Coulomb force, which means that the total complement of energy induced in a charged particle by the electric field related Coulomb force can be calculated with the following equation directly drawn from Equation (3.3):

$$K_{Total} = mv^2 = \frac{e^2}{4\pi\varepsilon_0 r} \tag{3.5}$$

which is the total amount of induced energy that Leibnitz already considered in Newton's era as being the real effect of application of a force ([57], p. 222).

3.4. Special Relativity and the gamma factor

As previously mentioned, the gamma factor on its part was first established by Woldemar Voigt in 1887 [72], for whom there are on record epistolary contacts with Larmor, Lorentz and Poincare, who also are credited with developing the method. This method is clearly laid out in a very well done paper published in 2003 by Richard E. Haskell [71]. Indeed, it is clear from this description, that the gamma factor was established strictly from formal geometric and trigonometric considerations unrelated to electromagnetism.

On page 10 of Reference [71], the first postulate of Special Relativity is summarized as resolving to the following statement "*Absolute uniform motion cannot be detected by any means.*" and the second postulate is formulated as "*Light is propagated in empty space with a velocity c which is independent of the motion of the source*".

It must be noted here that these postulates are presented as being axiomatic in nature, since they are not presented as deriving from underlying experimentally established physical causes.

It is useful also to note here that these postulates were proposed in this axiomatic manner by Einstein in 1905 without any mention of the fact that the constant velocity of light in vacuum and its known exact velocity of c=299792458 m/s had been established 40 years earlier by Maxwell from second partial derivatives of Gauss and Ampere equations, that linked both electric and magnetic fields as mutually inducing each other in a manner that could only result in this stable velocity of electromagnetic energy in vacuum.

On its part, the first postulate was briefly mathematically demonstrated by Poincaré grounded on the Lorentz transformation in a 4 pages note published in June of 1905 [76], shortly followed by a complete demonstration published in January 1906, titled "*Sur la dynamique de l'électron*" ("*On the Dynamics of the Electron*") [77], concluding that the Lorentz theory would completely explain the impossibility to demonstrate absolute motion, if all forces were of electromagnetic origin. So it appears that there was general consensus in the community at the time that both postulates were valid as Einstein grounded his Special theory of relativity on them, which explains why both his theories became so rapidly popular at the beginning of the past century.

It must thus be realized that the totally conclusive experimental verifications of the speed of electromagnetic energy in vacuum by a variety of means over the course of the past century indeed first and foremost validate Maxwell's calculations, which were arrived at not axiomatically, but were derived from equations experimentally established by Gauss and Ampere. Indeed, the constancy of the speed of light is so well established experimentally that in 1983, the SI meter was redefined as being the experimentally confirmed and fixed distance covered by light in 1 second divided by 299792458.

On its part, the alleged impossibility to demonstrate the absolute motion of the earth, as mathematically demonstrated from an analysis of the Lorentz transformation by Henri Poincaré [76] [77], is now strongly questioned by the discovery that the momentum energy of each elementary particle of which the earth is made physically exists and that it is strictly the easily measurable level at any given moment of this adiabatic energy that determines its state of absolute motion. This will be discussed in Section 3.5.1.

To relate the constancy of the speed of light as the foundation of the SR theory, the traditional procedure makes use of the famous relation between two different reference frames moving inertially at different constant velocities, each harboring an observer immobile in his own reference frame, both having the task of measuring the speed of a light beam to be the same for both observers.

As described on page 10 again, the traditional set up involves that one of the inertial reference frames be a train moving at a fixed velocity, and that if a light signal was emitted from the back of the train to the front, then both an observer on the train and an observer on the ground should be able to measure the velocity of the light signal as being c.

Then is exposed the logically well grounded geometric construct that allows associating with Equation (5) of Reference [71] a squared velocities ratio v^2/c^2 to the *sin* component in the well known trigonometric function $sin^2\ \theta + cos^2\ \theta = 1$, to then establish a precursor to the gamma factor as being related to time (but also axiomatically to the concept of time dilation) with Equation (6) of Reference [71]. It is quite interesting to note at this point that this particular trigonometric function can also be used to describe the mutual induction of the electric and magnetic fields of localized quanta such as electromagnetic photons ([15], [8] Chapter 6), as we will see further on.

It is to be noted also that c is axiomatically introduced in this relation without reference to its prior establishment from experimentally defined electromagnetic equations by Maxwell. The same procedure is then used to relate the same squared ratio of velocities to the *cos* component of the same trigonometric function to relate the length of the train (thus length contraction) to this other precursor of the gamma factor as Equation (8) of Reference [71].

The gamma factor is then formally established with Equation (14) of Reference [71] as being related to time dilation and length contraction of moving bodies. The remainder of Part II and Part III describe the Lorentz transformation and relativistic dynamics from the Special Relativity perspective.

3.5. Disconnect between the distance-dependence of energy induction and the concept of SR length-contraction

It is at this point that the distance-dependence of kinetic energy induction in charged particles by the Coulomb force as established with Equation (47-19) of Reference [57] previously reproduced as Equation (3.4) must be brought to mind again, because there is a clear disconnect between this property of the Coulomb force which is in permanent action between charged particles and the concept of length contraction as applied in Special Relativity to moving macroscopic bodies.

To correctly put this issue in perspective, it is important to become aware of the physical distances separating electronic escorts from nuclei within atoms. If for example a hydrogen atom was upsized so that the proton became as large as the Sun, then the electron would stabilize as far as Neptune's orbit, which would make the whole atom as large as the solar system! This means that all proportions considered, the distances separating electronic escorts from nuclei within atoms are relatively astronomical with respect to the sizes of elementary particles.

Given that all macroscopic bodies are made of such practically *empty* structures, the very concept of *length* becomes meaningless with respect to their internal composition, and what would be involved when the possible *length contraction* of a macroscopic body is considered, would really be a *distance contraction* between the electronic escorts and the nuclei of the constituting atoms, which is the only way that the physical length of a rigid macroscopic body can be diminished without deformation.

This being said, such distance contraction would apply by structure not only to the length of macroscopic bodies, but also to their other dimensions, which are their width and thickness, and such shortening of the distances between the charged electrons of the electronic escorts and their charged atomic nuclei within bodies subjected to *length contraction* would then involve by structure a corresponding energy increase within the mass of the body due to the now increased intensity of the Coulomb force at these shorter distances between the charges.

However, no such increase in energy is even considered in SR in relation to *length contraction* of moving macroscopic bodies, which means that despite the general assumption that SR is electromagnetism compliant, it really is not, because the Coulomb law is at the heart of Gauss's equation for the electric field, and is in fact Maxwell's first equation, from which the Coulomb Equation (3.2) can easily be derived ([32], [8] Chapter 14).

Another major issue can also be raised with regard to the relation between the gamma factor axiomatically established from strictly geometric and trigonometric considerations relating it to time dilation and length contraction, and its use to conclude that Maxwell's equations and Lorentz's force equation can be derived from SR as described in Part IV of Reference [71]. Since the gamma factor seems to never have been derived from an electromagnetic equation, such an interconnection of electromagnetism to time dilation and length contraction is at best axiomatic.

Indeed, despite a long and fruitless search in formal literature for such a derivation, evidence seems to reveal that a first time derivation of the gamma factor from an electromagnetic equation was effectively carried out and published only in 2013, as Equation (66) in Reference ([42], [8] Chapter 5), derived from Equation (51) of the same reference, itself a conversion from strictly electromagnetic Equation (34) of the same reference, and from which all relativistic equations can be derived ([42], [8] Chapter 5).

Equation (34) from Reference ([42], [8] Chapter 5) is indeed derived in direct line from the Biot-Savart equation via a seamless derivation by Paul Marmet directly relating the relativistic mass increase of a moving electron to a simultaneous increase of its magnetic field with velocity ([42], [8] Chapter 5) [29].

And even if a prior derivation of the gamma factor from an electromagnetic equation had been carried out while escaping the attention of this author, the outcome would be the same, because it can effectively be verified that from the electromagnetic perspective the *gamma factor* derived in Reference ([42], [8] Chapter 5) has nothing to do with time dilation nor length contraction, but is strictly related with charged particles momentum related kinetic energy increase with velocity and proximity between charged particles according to the Coulomb law, in accordance with Equation (3.5), and in conformity with the conclusions of Voigt, Abraham and Poincare with regard to Kaufmann's experiments [36] [39] [72] [75].

So, whatever dimensions may be associated to the varying ratio of the gamma factor, m/s, joules or kg, these dimensions always simplify completely out of whatever calculation the gamma factor may be involve in, which means that the Lorentz factor is only a special case of an intrinsically dimensionless mathematical function that can be used in a general manner to introduce the denominator of the ratio as the asymptotic limit of a growth curve obeying the power of the ratio, in the present case, the squared ratio and the asymptotic limit derived from an electromagnetic equation.

Consequently, there seems to be ample reasons to question the compliance of SR with Maxwell equations, and there is also reason to question the reality of time dilation and length contraction axiomatically related to the gamma factor as established from strictly geometric and trigonometric considerations when put in perspective with respect to the direct derivation of the same gamma factor from an electromagnetic equation that relates it strictly to the variation of energy induced by the Coulomb force as a function of the distances separating charged particles.

It goes without saying that such questioning of the reality of time dilation and length contraction axiomatically established as a foundation of SR also brings in question the space-time curvature of the General Relativity theory and all of the axiomatic conclusions that the theory leads to. It must be emphasized here that Einstein himself had become convinced toward the end of his life that gravitation follows the pattern of electromagnetism ([7], p. 391), which means that he also had come to doubt the validity of his own brainchildren SR and GR theories. See also Sections 1.7.1 and the Foreword about this issue.

These considerations are at the heart of the development of the present possible alternate solution entirely derived from the converging set of electromagnetic equations experimentally established by Coulomb, Ampere, Gauss, Faraday, Maxwell, Lorentz, Biot and Savart, without any axiomatic assumptions.

One of its main objectives was to attempt addressing one of the major hurdles of fundamental physics which is summarized in this remark by Feynman mentioned during his famous *"Feynman Lectures on Physics"* [26]:

> *"There are difficulties associated with the ideas of Maxwell's theory which are not solved by and not directly associated with quantum mechanics...when electromagnetism is joined to quantum mechanics, the difficulties remain"*

This author is convinced that by clearly defining the self-sustaining mutual induction of the electric and magnetic fields of the energy quanta making up localized electromagnetic elementary particles such as the electromagnetic photon and the electron, this hurdle will be resolved.

Permanent localization of the electron when in motion is maintained in this new paradigm by allowing the definition of a clear resonance trajectory of the moving electron within the volume defined by the wave function.

3.5.1. Relative Reference Frames and Absolute Motion

This analysis of the foundation of the SR theory now brings up the old issue of the role played by the use of relative reference frames to draw conclusions in fundamental physics, that was popularized by Einstein's famous thought experiment previously mentioned, involving two different reference frames moving inertially at different constant velocities, one being materialized as a train in which an observer is moving at the same velocity as the train, and the other being materialized as a stationary observer standing on the ground as the train is moving by.

The conclusions drawn from this thought experiment, are what completely underlies the traditional concept of *relativity*, that led to the elaboration of the

Special Relativity theory, that is, the definition of *relative motion* proposed by Poincaré [76] as perceived by *observers*, but not with respect to data collected from physically carried out experiment, as was the case for the establishment of the set of mutually converging electromagnetic equations that successfully underlies all of the technological achievements that we currently benefit from.

It goes without saying that such inertial reference frames moving at fixed velocities can only be idealized mathematical concepts, since it is impossible for such fixed velocities to naturally occur in physical reality, whence the difficulty involved in matching conclusions drawn from such idealized thought experiments with really occurring physical processes.

The transposition from using the gamma factor as an idealized mathematical concept as defined by Voigt, leading to the observed disconnect between the *idealized concept of length contraction* of masses and the *physically existing distances* between elementary particles within the atoms of which these real macroscopic masses are made, that was analyzed in Section 3.5; to using it as being derived from an electromagnetic equation ([42], [8] Chapter 5), itself derived from a chain of equations initially established from the analysis of data collected during physically carried out experiments, that relates it rather to the manner in which energy is adiabatically induced in elementary charged particles by the Coulomb interaction, definitely highlights the benefits of grounding theories on equations emerging from the analysis of repeatably obtainable data gathered from physically carried out experiments rather than on idealized axiomatic premises.

This habit of hypothesizing inertial reference frames in countless other attempts to make sense of the experimentally obtained data then became deeply ingrained in the community since the beginning of the 20th century. The fundamental question that this method was meant to address is the following:

What is the motion of masses relative to in physical reality?

Is it relative to some underlying medium? To the point of emission? To the point of absorption? To the observer? To this or that reference frame, or multiple reference frames, inertial, non inertial, Galilean, moving or not? etc.?

The century old line of research and development of concepts of motion *relative to observers* with all of its complexities has its roots in the established certainty that it is impossible to demonstrate *absolute motion* in the universe.

But the manner in which the present analysis reveals that kinetic energy is adiabatically induced in all charged particles leads to observe that they all can only be self-propelling according to the amount of ΔK momentum energy that they physically possess. So their motion, thus their velocity, can depend on only one criterion, which is the actual presence of their translational momentum ΔK kinetic energy component. As analyzed in Reference ([15], [8] Chapter 6), if the local electromagnetic equilibrium state allows it, there will mandatorily be a velocity of the particle expressed in vacuum, irrespective of any hypothesized reference frame or frames.

Consequently, motion in the universe is relative only to the constantly measurable amount of momentum energy that each charged particle locally possesses (in its own reference frame) at any given instant.

Traditionally, it is assumed that in its own reference frame an elementary particle such as an electron or a photon has no velocity, which implies, according to the classical concepts, that its momentum would fall to zero. But the current analysis confirms that from the electromagnetic perspective, the ΔK and Δm_m *carrying energy* components of the particle have permanent existence as a *physically existing substance*, which means that from the very particle's structure and its continuously varying amount of its ΔK and Δm_m energy components, the state of *absolute motion* of the particle can be continuously determined within its own inertial reference frame, with the trispatial energy-momentum Equation (1.50) (See Appendix A) for both a massive elementary particle such as the electron, for example, and by setting m_o to zero in the equation, also for a free moving photon:

$$E_e = \Delta K + \Delta m_m c^2 + m_0 c^2 \tag{1.50}$$

From these energy values, its velocity can then be calculated with either Equations (1.33), and this, strictly from information available in the particle's own reference frame:

$$v = c \frac{\sqrt{\lambda_c(4\lambda + \lambda_c)}}{(2\lambda + \lambda_c)} \quad \text{or} \quad v = c \frac{\sqrt{4EK + K^2}}{2E + K} \tag{1.33}$$

Moreover, from within the particle's reference frame, the variation over time of the total amount of an electron's carrying energy will reveal its state of *absolute motion* with respect to its environment, and therefore *its state of absolute motion in the universe.* Rapid increases and decreases of it total amount of energy will reveal that it is stabilized in some stationary action resonance state. Slow increase and decrease of its maximum amount of carrying energy over longer periods of time will reveal that it belongs to an atom which is part of a macroscopic mass stabilized on an elliptic macroscopic orbit in some planetary system, and so on.

The <u>absolute lower velocity limit</u>, as seen from this perspective, would be an electron possessing zero translational kinetic energy in excess the energy making up its rest mass. But such an electron totally deprived of translational kinetic energy can only be theoretical since all charged particles are subject to electrodynamic acceleration in physical reality from the moment they start existing, and it is then not physically possible for them not to be induced with some amount of carrying energy by the ambient Coulomb interaction.

The <u>absolute upper velocity limit</u> involving electromagnetic oscillation is reached when a ΔK amount of translational kinetic energy propels an equal amount of Δm_m kinetic energy captive in transverse electromagnetic oscillation, that is, a free moving electromagnetic photon. Its well known velocity of c is the speed of light, that can be calculated with the second Equation (1.33), if E, the energy making up the invariant rest mass of the carried charged particle, is set to zero. See Equations (2.43) and (2.44).

The <u>only other possible case between these two limits</u> involving electromagnetic oscillation, applies to an amount of kinetic energy captive in transverse electromagnetic oscillation being propelled by a lesser amount of translational kinetic energy, such as the kinetic energy making up the $m_o c^2$ rest mass of an electron, plus the transversely oscillating Δm_m half of its carrier-photon's kinetic energy, both quantities being propelled by the ΔK unidirectional half of the carrier-photon's quantum of kinetic energy. The velocity of such a system will mandatorily lie between zero and asymptotically close to the speed of light, a process whose mechanics is described in Reference ([42], [8] Chapter 5), and can be calculated with both Equations (1.33).

Finally, the remaining case of kinetic energy whose motion seems not to involve any transverse electromagnetic oscillation and for which there consequently also seems not be any limiting factor on the velocity is the case of escaping neutrino energy, whose mechanics of liberation in the trispatial model is described in Reference ([33], [8] Chapter 12).

3.6. Establishment of the fundamental equations from physically collected data

One of the major difficulties in fundamental physics is the very power of mathematics as a descriptive language. If care is not taken to avoid as much as possible axiomatic postulates, an indefinite number of theories can be elaborated with full mathematical support that can always become entirely self-consistent with respect to the set of premises from which each theory is grounded. But the very self-consistency of all well thought out theories is so appealing to our rational minds that it renders very difficult the requestioning of the grounding foundations of such beautiful and intellectually satisfying structures and consequently the identification of possibly inappropriate axiomatic assumption.

Given however that there exists only one physical reality, it would seem that only one explanation would correctly address each of its aspects and that the related theories are likely to more appropriately describe it if axiomatic assumptions are avoided as much as possible in grounding their elaboration.

Before drawing any conclusion about physical reality, experimental data about this physical reality must obviously be first collected, that then allows extrapolating hypotheses that would explain this data. Although it is relatively easy to confirm the validity of this data by repeatedly obtaining the same results from various experimental means, the same cannot be said of the theories established from the interpretation of this data.

For example, the unit charge of the electron has been conclusively measured out of any possible doubt over the course of the past century, to have an absolutely invariant value. Therefore, this characteristic of the electron is considered an objectively valid component of any set of premises to be used in drawing conclusions about its nature. However, the conclusion as to whether the electron remains localized while moving, as dealt with from the relativistic mechanics perspective, or whether its *substance* spreads out when moving according to the wave function, as dealt with from the quantum mechanics perspective depends entirely on the other elements in the set of premises on which each theory is grounded.

Other conclusively confirmed characteristics of the electron are the invariance of its rest mass, the fact that it always behaves point-like during scattering encounters and that it presumably possesses an indefinite life span unless physically converted to energy during very specific accidental interaction events with other elementary particles, which allows considering it as being *stable*.

Since all matter in existence is made of massive atoms, gravitation logically must be emergent from the properties of these atoms. In turn, all atoms being ultimately made of a very restricted set of charged and massive elementary particles locked in mutual interaction, this logically implies that the properties of atoms must ultimately be emergent from the properties of these elementary subcomponents.

This ultimate set of stable, charged and massive elementary particles making up the internal structure of all atoms is very limited, and their existence has been confirmed out of any doubt by means of non-destructive scattering. There are only 3 of them, that is, the electron, that establishes the electronic escorts of atoms about their nuclei, which determines atomic volumes; and the up and down quarks that were found to be the ultimate charged and massive elementary subcomponents of all nucleons inside atomic nuclei, that determine their volume, and that were first detected by deep non-destructive scattering during the first years of operation of the Stanford Linear accelerator (SLAC), from 1966 to 1968 [21].

These three charged particles are considered elementary because no scattering experiment ever revealed the existence an unbreachable limit at some distance from their center that would have related them to a measurable volume, as was the case for protons and neutrons, when scattered against with insufficient energy, which was the unmistakable telltale that nucleons are not elementary and have an internal structure involving smaller particles, that turned out to be the up and down quarks just mentioned, observed to be interacting in triads of both types, that is, *uud* for the proton and *udd* for the neutron. See Sections 1.23 and 2.16.

These three particles being elementary, their masses must then logically be made of some undifferentiated substance, which was identified in the case of the electron as being electromagnetic energy, given its electric and magnetic properties, and the same conclusion can be reached by similarity for the up and down quarks for the same reason.

We also know that electromagnetic energy is intimately linked to momentum related kinetic energy, because we have undisputable evidence that the exact amounts of kinetic energy accumulated by electrons accelerating between the electrodes of a Coolidge tube, for example, due to the Coulomb force in action between the accelerating negatively charged electrons and the positively ionized atoms of the anode, are liberated as electromagnetic X-ray photons when they are suddenly stopped in their translational motion, when momentarily captured by the positively ionized atoms of the anode (or anti-cathode).

We thus observe that from the electromagnetism perspective, the Coulomb force, that we know to be in continuous action between all charged particles in existence, belongs to the deepest layer of physical reality with regard to the induction of kinetic energy in accelerating charged elementary particles. Therefore, this force can be identified as the ultimate cause the very existence of kinetic energy at the submicroscopic level. We also know from the experimental evidence provided by the Coolidge tube operation, that as it escapes as Bremsstrahlung photons, this induced kinetic energy displays the same electromagnetic properties already associated with the restricted set of the three charged and massive electromagnetic elementary particles that are the only building blocks of all atoms in existence.

3.7. Procedure

This article will first put in perspective an aspect of electromagnetic energy yet unclarified in all currently useful physics theories, which is the fact that none of these theories provides a mechanical description of the self-sustaining mutual induction of the electric and magnetic fields of the energy making up the rest mass of elementary particles, that would be consistent with their point-like localization observed during their mutual scattering encounters, which is the mutual induction of electric and magnetic fields that justifies the very existence of electromagnetic energy in Maxwell's theory.

A possible description of this mutual induction in the frame of an expanded orthogonal space geometry will be proposed that brings to light a set of properties that allows mechanically explaining electronic and nucleonic orbitals stability in atoms. Such a mechanics is proposed in Section 2.20.

The function of the Coulomb force in electromagnetic energy induction will be analyzed, and a related analysis will follow of the disconnect that this new perspective reveals between the current energy conservation principle based concepts of momentum / Lagrangian / Hamiltonian and the currently unaccounted for motion hindered adiabatic kinetic energy permanently induced in these three charged elementary particles captive in various resonance states within atomic and nucleon structures.

The relation between these least action electromagnetic resonance states and the wave function as well as with gravitation will finally be put in perspective, as well as the possibility brought to light that the methods of quantum mechanics could be directly applied to nucleons inner structures.

3.8. The inner electromagnetic structure of electrons

One characteristic of the electron yet to be mentioned, was suspected by Louis de Broglie in the 1920's, and was experimentally confirmed in the 1930's. It is the fact that the very substance of which its invariant rest mass is made is actually electromagnetic energy, as established by the repeatedly confirmed fact, initially discovered by Blackett and Occhialini [78], that massless electromagnetic photons of 1.022 MeV or more can be destabilized into converting to massive electron-positron pairs and that the masses of a pair of electron and positron metastabilizing into positronium configuration will reconvert back to massless electromagnetic photon state as the final inward spiraling stage of the positronium decay process, which was also initially confirmed by Blackett and Occhialini in the same period. Confirming evidence of the electromagnetic nature of the mass of these two particles is of course that they are electrically charged and are conclusively known to possess a magnetic moment.

These intrinsic electromagnetic properties of the energy constituting the rest mass of the electron are however not clearly integrated into its wave function representations, nor are they integrated into the concept of localized mass addressed by relativistic mechanics.

3.9. No description of the electron internal electromagnetic structure in classical and relativistic mechanics

Relativistic mechanics treats all massive bodies, including electrons, as if they had no internal structure, which sometimes leads to results that are difficult to relate to otherwise well established laws.

For example, in traditional classical and relativistic mechanics, this difficulty is made particularly obvious with respect to the rotating motion of massive bodies, whose angular momentum is deemed to be conservative, which is a conclusion that disregards the fact that in physical reality, all macroscopic rotating masses can only be made of the sum of the masses of a number of captive elementary massive particles translating on circular orbits about the body's axis of rotation, all of which are individually subject to the 2nd principle of thermodynamics, that mandates that the constant change in direction imposed on these massive sub-components *de facto* involves an expenditure of energy as work, which comes in contradiction with defining the rotation of macroscopic bodies as being conservative, given that it is impossible, according to the 2nd principle of thermodynamics, that the state of motion of massive bodies such as these massive elementary particles could constantly change in this manner without an expenditure of energy.

Could this be related to the unexplained observed rotation slowing down of all bodies left to rotate in deep vacuum for extended periods of time after having been set in rotation from an initial impulse, such as both Pioneer 10 and 11 spacecrafts ([79], p. 23)? Or the ball bearing of J.C. Keith's experiment in 1963 [80], that was made to rotate frictionlessly at high velocities in deep vacuum while suspended in magnetic fields? Or the ball bearing in a similar confirming experiment conducted by J.K. Fremerey in 1973 [81]? Or even of individual electrons being made to translate on perfectly circular orbits in the Betatron during J.P. Blewett's experiments in 1946 ([82], p. 87)?

Unfortunately, the case of the still unexplained electron translational slowdown observed by Blewett was not investigated further in a Betatron accelerator and was neglectfully left hanging without an answer to this day, after the G.E. 100 MeV Betatron was decommissioned before he could investigate further. Various reassuring rationalizations were applied to all other slowdown cases that did not force reconsideration of the assumed *conservative* nature of rotating motion.

In CERN Reference [83], the case of the excess energy that needed to be constantly provided in the GE Betatron for electrons to be maintained on their assumed *conservative* perfectly circular trajectory, without any radiation being detected to explain the excess energy required, was rationalized in this manner:

> *"Then in 1946 Blewett measured the energy loss due to SR in the G.E. 100 MeV Betatron and found agreement with theory, but failed to detect the radiation after searching in the microwave region of the spectrum [4]. It was later pointed out by Schwinger however that the spectrum peaks at a much higher harmonic of the orbit frequency and that the power in the microwave region is negligible [5]."* ([83], p. 463)

The case of the *lost energy* measured by Blewett was then considered settled by the first observation of *synchrotron radiation* in the G.E. 70 MeV Synchrotron that entered service in 1947:

> *"The first direct observation of the radiation as a "small spot of brilliant white light" occurred by chance in the following year at the G.E. 70 MeV synchrotron, when a technician was looking into the transparent vacuum chamber [6]."* ([83], p. 463)

However, it can be forcefully asserted that the case is not settled, because synchrotrons by very design are unable to exactly reproduce a *perfect circular orbits* with respect to which Blewett made his measurements and that can be realized only in a Betatron type accelerator, because the slightest transverse oscillation out of perfect circular orbit will obviously generate transverse bremsstrahlung radiation in a synchrotron, however *close* to a perfectly circular orbit electrons can be constrained into in a synchrotron. Schwinger's argument may be valid of course, but a really conclusive test can only be carried out in a Betatron, which is the only existing accelerator design that, by structure, is able to force isolated electrons to move on a *perfectly circular orbit* on which a loss of energy was observed *without energy emission being detected*.

Consequently, an experiment bringing an actual confirmation of *energy loss without radiation during perfectly circular motion not naturally compensated* ([52], [10] Chapter 10) for elementary electromagnetic particles as measured by Blewett remains to be carried out.

It can effectively be observed that all elementary massive particles captive inside macroscopic rotating bodies are translating on perfectly circular macroscopic scale orbits identical to that of the isolated electrons observed by Blewett in the Betatron. It would be highly interesting indeed if the still to be investigated unexplained slowdown of the Betatron electrons was finally studied in depth and correlated with the slowdown observed for rotating bodies, taking into consideration the identical circular orbits that elementary charged and massive particles are forced into within macroscopic bodies.

Interestingly, it wouldn't come to anybody's mind to consider the solar system as being a massive body devoid of an internal structure, because we can directly observe that it is a stabilized system of smaller bodies and that its total mass is the sum of the masses of these individual smaller bodies. This is nevertheless what is being done when not assuming an internal structure to macroscopic massive bodies, because it is not the macroscopic bodies themselves that are massive, but the individual submicroscopic elementary massive particles whose sum of individual masses add up to make up the total mass of any macroscopic massive body.

Not assuming an internal structure for macroscopic masses, classical and relativistic mechanics do not assume any internal structure either for the rest mass of elementary electromagnetic massive particles such as the electron, which led to the default unexpressed assumption that the electron has such a *volume*, simply by finding no fault with the concept of magnetic spin as corresponding to an angular momentum, since the very concept of *rotation* mandates the presence of such a volume, which disregards the fact already mentioned that no scattering experiment ever detected any unbreachable limit at some distance from the point-like center of electrons that would have related them to such a measurable volume, as was the case for protons and neutrons.

Indeed, the only logically possible cyclic process that could animate a volumeless object such as the electron, that behaves point-like during all scattering encounters, seems to be some sort of reciprocating internal motion, a hypothesis that will be supported by the manner in which the self-sustaining internal mutual induction of the electric and magnetic fields of the electromagnetic energy quantum making up its invariant rest mass can be represented in an expanded orthogonal space geometry, as will be shown further on.

3.10. No description of the electron internal electromagnetic structure in quantum mechanics

Quantum mechanics on its part currently offers three different descriptions of the electron in motion, which comprises the energy making up its rest mass plus its momentum energy, but does not offer separate representations of these two quantities.

The first representation stems from Schrödinger's wave function description that he established to represent the resonance states that de Broglie had previously concluded that electrons had to be captive into when stabilized about atomic nuclei [4]. Stated in general terms, this representation describes the electron energy as being spread out within the volumes definable with the wave function.

The second representation was simultaneously and independently developed by Heisenberg, which spreads the energy of the electron within the volume otherwise defined by the wave function according to statistical probabilities of density presence of the energy of which the electron is made, which allows for example defining the area of greatest probable density of the electron *substance* in the hydrogen ground state as corresponding to the ground state orbit of the classical Bohr atom.

The third representation is the path integral subsequently developed by Feynman, which replaces the theoretical least action trajectory of an electron in motion with the infinity of possible trajectories that the electron could possibly run within the volume defined by the wave function.

It can also be observed that besides not representing separately the carrying energy of the electron from its invariant rest mass, none of these current QM representations offers a description of the self-sustaining mutual electric and magnetic induction of the energy quantum making up its invariant rest mass.

We will see further on how a possible fourth representation, that makes use of such a description may allow describing the actual resonance trajectory of a permanently localized electron within the volume defined by the wave function in the ground state of the hydrogen atom, thus proposing a possible general method to allow resonance trajectories representation of localized charged elementary particles within all atomic and nuclear orbitals within the limiting volumes definable by the wave function. This fourth representation is completely analyzed in Sections 2.17 to 2.20.

3.11. No description of elementary particles internal electromagnetic structure in quantum field theory

The more general quantum field theory (QFT) assumes that elementary electromagnetic particles such as the electron emerge as *local excited states* of an underlying neutral quantum energy field, thus introducing the concept of energy quantization that gave rise to quantum electrodynamics (QED), which allows the description of interactions between elementary particles as quantized *virtual exchange photons*.

But it can be noted that QFT, although grounded on electromagnetism, provides no description either of the actual internal mutual induction process of both electric and magnetic fields of these individual excited states.

3.11.1 Progress also resuming from the QFT perspective

As mentioned at the very beginning of Chapter 1, Quantum Field Theory emerges from Ludwig Lorenz's interpretation of the manner in which the electric and magnetic fields of free moving electromagnetic energy have to relate to each other to explain the velocity of light.

As put in perspective already, Lorenz considered that the E and B fields of electromagnetic energy have to synchronously peak at maximum at the same time for this velocity to be maintained, while Maxwell considered that both fields had to mutually induce each other cyclically for the velocity of light to be maintained, both interpretations being equally consistent with the complete electromagnetic equations set.

Although Maxwell's interpretation leads to an electromagnetic mechanics allowing the description of the complete sequence of interactions between electromagnetic elementary particles as continuously progressive sequences, there is no doubt that an equivalent electromagnetic mechanics that would describe these sequences as being discontinuously progressive can also be developed, that would also take into account the transverse magnetic component of electromagnetic quanta, given the success already achieved by Quantum Electrodynamics (QED), that already emerges from QFT, and that currently puts more emphasis on the electric aspects of these interactions.

It so happens that electrical engineers deeply knowledgeable in electromagnetism, Riccardo Storti and Todd Desiato, have already been clarifying QFT further recently [84]. All considered, how unsurprising is it that it would be engineers who are now pushing the limits in fundamental physics, considering that rigorous mathematical competence is not an option in their field, and that the true repository of humankind's applied and applicable knowledge about all aspects of mechanics and electromagnetics is precisely the set of engineering Reference works, as put in perspective in Reference ([25] Section 27), such as References [85] [86] [87] [88] [89] [90] to give only a few examples, experimental data repositories such as References [35] [51] and [73], and outstandingly well made introductory textbooks such as References [17] [18] [57] [62] [63].

3.12. No description of the electron internal electromagnetic structure in electromagnetism

Surprisingly, even in electromagnetism as currently formulated, although the very foundation of Maxwell's theory mandates that the electric and magnetic fields of free moving electromagnetic energy need to cyclically induce each other for electromagnetic energy to even exist, it has not proved possible to this day to coherently represent this cyclic self-sustaining mutual induction within localized electromagnetic photons, nor within localized elementary electromagnetic particles such as the electron.

Indeed, it was the observation that a mechanical description of this internal mutual electric and magnetic fields induction was lacking in all of these generally useful theories about matter and energy that brought to light the possibility that resolving this particular issue might clarify some aspects of electromagnetic energy that could be key to completely reconcile these theories with each other and with objective reality.

As quantum mechanics was being established in the 1920's, it was already obvious of course that electromagnetism needed to be correlated with the newly developed wave function due to the prior establishment by H.A. Lorentz of the groundbreaking first ever equation of electromagnetic mechanics $F=q(E+v \times B)$, that allowed controlling the motion of electrons on precise trajectories by varying the relative densities of ambient electric and magnetic fields, equal densities providing motion in straight line of the charged particle.

So, very early after the advent of Schrödinger's wave function and Heisenberg's statistical method, this disconnect between QM and electromagnetism eventually gave rise to the development of QFT, which led to the introduction of the quantization perspective.

Louis de Broglie on the other hand, remained intimately convinced that electromagnetic photons and electrons had to remain permanently localized and follow precise trajectories when in motion. He then undertook to establish the internal electromagnetic structure for the localized electromagnetic photon [91] [92] [93] [94], but his decade long attempt in the 1930's not succeeding in correlating this internal mutual induction in harmony with the wave function, led him to conclude that it was impossible to exactly represent elementary particles in the frame of the 4 dimensional space-time geometry, also adding that this should eventually become possible by escaping from this presumably too restrictive space time frame ([27], p. 273).

Even today, we know that light can be polarized, but we do not have a mechanical description explaining why electromagnetic energy can be polarized.

Even if we know that electromagnetic energy involves a process of mutual electric and magnetic fields induction, we still do not have a mechanical description explaining why the electromagnetic energy quantum that makes up the rest mass of an electromagnetic elementary particle such as the electron can remain localized while its internal electric and magnetic fields mutually induce each other in the self-sustaining manner that we can observe.

We know that only three stable, charged and massive electromagnetic elementary particles are the only building blocks of all atoms in the universe (electron, up quark and down quark), but we do not know yet why their electromagnetic energy quanta, made of self-sustaining and mutually inducing electric and magnetic fields, remain localized to display the point-like behavior that we can observe when they scatter against each other.

Since gravity is apparently related to mass, it must by structure be related to the only three existing stable self-sustaining electromagnetic elementary particles that display mass and that obviously are the only and ultimate massive building blocks of all existing atoms in the universe.

3.13. Establishing the internal structure of electromagnetic photons

Following de Broglie's intuition that 4D spacetime geometry seemed too restrictive to establish this mechanics, a new expanded space geometry was eventually developed and proposed in 2000 [28], that relates the triply orthogonal relation of electromagnetic energy revealed by plane wave treatment with the orthogonality of space itself, that effectively allows the mechanical representation as Equation (3.6) of the internal mutual induction of the electric and magnetic aspects of the electromagnetic energy of a localized photon as hypothesized by de Broglie, in conformity with Maxwell's equations, in a manner that can explain polarization ([15], [8] Chapter 6). This new expanded space geometry is put in perspective with respect to other more familiar multidimensional attempts at resolving the remaining issues in fundamental physics in Reference ([34], [8] Chapter 19), and is completely described in Reference ([15], [8] Chapter 6).

Summarized in a few words, this expanded space geometry stems directly from the well known triple orthogonal vectorial relation of electromagnetism that maps any point of the wavefront of Maxwell's continuous electromagnetic wave as a cross product of the magnetic field vs the electric field, both perpendicular to the direction of motion of any point of the wavefront in plane wave treatment. The new geometry results from metaphorically *exploding* each of the three related mutually orthogonal *ijk* vectors into full blown 3D vectorial spaces mutually orthogonal to each other, while the central junction point of all unit vectors of such vector complexes remains located at the center of each localized electromagnetic quantum.

For example, the following trispatial LC Equation (3.6) for the self-propelling localized electromagnetic photon clearly describes in this expanded space geometry, how half of its energy transversely oscillates between a state of two electric components, which is key to explaining polarization in conformity with de Broglie's hypothesis, and a single magnetic state, that insures permanent localization of the quantum in complete conformity with Maxwell's equations; while the other half remains unidirectional, perpendicularly to the transversely oscillating half, sustaining the momentum related equilibrium speed of light of the whole quantum in vacuum, without any need for an underlying aether, while its default equal electric and magnetic fields densities ensure self-guidance in straight line when no external electromagnetic fields modifies this default equal densities equilibrium ratio in a manner that would deflect its trajectory ([15], [8] Chapter 6):

$$E\,\vec{I}\,\vec{i} = \left(\frac{hc}{2\lambda}\right)_X \vec{I}\,\vec{i} + \left[\begin{array}{l} 2\left(\dfrac{e^2}{4C}\right)_Y (\,\vec{J}\,\vec{j}\,,\vec{J}\,\overleftarrow{j}\,)\cos^2(\omega t) \\[2ex] + \left(\dfrac{L\,i^2}{2}\right)_Z \overleftrightarrow{K}\,\sin^2(\omega t) \end{array}\right] \tag{3.6}$$

where

$$C = 2\varepsilon_0\alpha\lambda \qquad L = \frac{\mu_0\alpha\lambda}{8\pi^2} \qquad i = \frac{2\pi\,ec}{\alpha\lambda} \qquad \omega = \frac{2\pi\,c}{\alpha\lambda} \tag{3.7}$$

3.14. Establishing the internal electromagnetic structure of the carrying energy of massive elementary particles

Regarding a possible representation of the internal electromagnetic structure of the energy of the rest mass of the electron with respect to relativistic mechanics, a groundbreaking breakthrough was made by Paul Marmet in 2003, when he succeeded in deriving from the Biot-Savart equation a relation that directly relates the relativistic mass increase of an accelerating electron to the simultaneous increase of its magnetic field [29], which led to observe that the magnetic field of the electron at rest corresponds to precisely half of its invariant rest mass, which in turn led to conclude that the other half of its invariant rest mass had to correspond to its electric field.

This means that by structure, the electron measurable velocity related relativistic magnetic mass increment can only involve its carrying energy as being distinct from the energy making up its invariant rest mass energy quantum, in a manner that causes it to acquire the same magnetic mass characteristics as the invariant rest mass of the electron ([30], [8] Chapter 4), that is, a property of omnidirectional inertia in normal space corresponding to the established concept of electromagnetic mass.

Indeed, since the traditional momentum related kinetic energy that propels the electron can only be vectorially unidirectional by structure as it translationally propels the electron, and that in accordance with Maxwell equations' fundamental vectorial requirement that a magnetic field be by structure oriented perpendicularly to the direction of motion of electromagnetic energy, then the magnetic field of the velocity related mass increment contributed by the energy in excess of the electron invariant rest mass, can only be part of a transversely oriented energy component different from the translationally oriented momentum energy component of the carrying energy, which thus exists by structure separately from the particle invariant rest mass electromagnetic energy quantum:

$$E_{\text{total carrying energy}} = E_{\text{translational}} + E_{\text{transversely oriented magnetic mass component}}$$

$$(3.8)$$

Given that a magnetic field cannot be dissociated from an electric counterpart in Maxwell's theory, and that both aspects mandatorily have to mutually induce each other for electromagnetic energy to even exist, then the only manner possible for this electric aspect to be introduced is for the transverse magnetic mass increment component of the carrying energy to be involved in a reciprocating swing, so to speak, between this magnetic state and a corresponding electric state:

$$E_{total} = E_{trans.} + \left[E_{elec.}\cos^2(\omega t) + E_{mag.}\sin^2(\omega t) \right] \qquad (3.9)$$

This form of the relation then obviously leads to the following LC representation:

$$E = \frac{hc}{2\lambda} + \left[\frac{e^2}{2C_\lambda}\cos^2(\omega t) + \frac{L_\lambda\, i_\lambda^{\,2}}{2}\sin^2(\omega t) \right] \qquad (3.10)$$

where

$$E_{\text{E(max)}} = \frac{e^2}{2C} \quad \text{and} \quad E_{\text{B(max)}} = \frac{L\, i^2}{2} \qquad (3.11)$$

There was then no failing to notice the similarity between Equation (3.10) stemming from the Biot-Savart equation and Equation (3.6) corresponding to the representation of a localized electromagnetic photon in the trispatial geometry, which led to the conclusion that the carrying energy of a moving electron can only have the same electromagnetic inner structure as that of a free moving localized electromagnetic photon; which allowed restructuring Equation (3.10) to incorporate the local electric and magnetic fields of the electron carrying energy as Equation (3.14) further on, which can indifferently be applied to the carrying energy of massive elementary particles and to free moving electromagnetic photons in replacement of Equation (3.6).

Further analysis then allowed mathematically demonstrating that the reason why the velocity of the electron *carrier-photon* was limited to velocities below the speed of light was uniquely due to the fact that the carrier-photon's momentum related unidirectionally oriented energy half is forced to propel the translationally inert electromagnetic invariant rest mass of the electron in addition to simultaneously propelling its own translationally inert electromagnetic other half, which can only slow it down accordingly, since the velocity of light of an electromagnetic photon is maintained in vacuum in this space geometry only due to the fact that it can only be made by structure of two equal halves, one of which remaining unidirectional while propelling the other half, which is translationally inert while electromagnetically oscillating transversely to the direction of motion, as analyzed in Reference ([42], [8] Chapter 5).

Finally, the fact that the relativistic mass increment of an electron in motion corresponds exactly to the transversely oscillating electromagnetic half of the electron carrying energy as represented by Equation (3.10), and that this relativistic mass increment possesses omnidirectional inertia just like the invariant rest mass of the electron, then allowed associating the same omnidirectional inertia properties to the transversely oscillating electromagnetic half of any free moving electromagnetic photon as represented by Equations (3.6) and (3.14), which provides a direct explanation to the observed deflection angle of light grazing stellar masses, without the need to resort to the curved spacetime solution of general relativity ([15], [8] Chapter 6) ([45], [8] Chapter 16).

3.15. *Establishing the internal electromagnetic structure of the rest mass of localized elementary particles*

From the method used by Marmet to derive his conclusion from the Biot-Savart equation, a new alternate general equation for calculating the energy of electromagnetic quanta equivalent to $E=h\nu$ was then derived that does not involve Planck's constant, and is obtained by spherically integrating their energy from infinity to a lower limit that relates their longitudinal wavelength to the fine structure constant; a lower limit corresponding to the transverse amplitude

($\lambda/2$) of the electromagnetic oscillation of a localized photon's energy quantum in the trispatial geometry ([30], [8] Chapter 4, Equation (11)):

$$E = hv = \frac{e^2}{2\varepsilon_0 \alpha \lambda} \tag{3.12}$$

This definition of energy incidentally allows observing that Planck's quantum of action belongs to a set of electromagnetic constants that inextricably define each other: $h=(e^2/2\varepsilon_0\alpha c)=(e^2\mu_0 c/2\alpha)$. Just as in the cases of Euler's identity $(e^{i\pi}+1=0)$ and the derivation revealing the speed of light in vacuum from Maxwell's equations $(\varepsilon_0\mu_0 c^2=1)$, it can be observed that the same logical rule to the effect that any value which is uniquely defined by a set of constants can only be itself a constant, now guarantees the invariance of Planck's constant.

In the present case, Planck's quantum of action can be confirmed being an electromagnetic constant by first confirming the invariance of the fine structure constant ([34], [8] Chapter 19, Equation (1)) with respect to three other previously established electromagnetic constants (e, ε_0, and also H, which is a newly defined electromagnetic intensity constant ([53], [8] Chapter 6, Equation (17))), and then confirming the invariance of Planck's quantum of action h ([34], [8] Chapter 19, Equation (4)) with respect to three previously established electromagnetic constants (e, ε_0, and α now confirmed to be invariant).

This new definition of energy in turn allowed defining the electric and magnetic fields of any localized photon, or massive elementary particle's carrying energy, from their wavelength and a specific set of known electromagnetic constants ([30], [8] Chapter 4):

$$\mathbf{B}=\frac{\mu_0 \pi e c}{\alpha^3 \lambda^2} \qquad \text{and} \qquad \mathbf{E}=\frac{\pi e}{\varepsilon_0 \alpha^3 \lambda^2} \tag{3.13}$$

Equations (3.13) applying equally to the energy quantum of free moving photons and to that of massive elementary particles' carrying energy, then allowed adapting Equation (3.6) to use these fields definitions instead of the less familiar and less handy capacitance and inductance definitions of Equations (3.7) to represent the internal electromagnetic structure of localized photons and also that of massive elementary particles' carrying energy:

$$E\vec{I}\vec{i} = \left(\frac{hc}{2\lambda}\right)_X \vec{I}\vec{i} + \left[2\left(\frac{\varepsilon_0 \mathbf{E}^2}{4}\right)_Y (\vec{J}\vec{j},\vec{J}\overleftarrow{j})\cos^2(\omega t) + \left(\frac{\mathbf{B}^2}{2\mu_0}\right)_Z \overleftrightarrow{K}\sin^2(\omega t) \right] V \tag{3.14}$$

where

$$V = \frac{\alpha^5}{2\pi^2}\lambda^3 \tag{3.15}$$

Equation (3.15) determining the volume that must be associated with fields Equations (3.13) to implement Equation (3.14) is drawn from a deep analysis carried out in Reference ([30], [8] Chapter 4) of the equation giving the energy density associated with electric field Equation (3.13) when the density of both E and B fields are equal in the context of straight line motion of charged electromagnetic elementary particles:

$$U = \varepsilon_0 E^2 = \varepsilon_0 \left(\frac{\pi e}{\varepsilon_0 \alpha^3 \lambda^2} \right)^2 = \frac{\pi^2 e^2}{\varepsilon_0 \alpha^6 \lambda^4}$$
$$= \frac{e^2}{2\varepsilon_0 \alpha \lambda} \times \frac{2\pi^2}{\alpha^5 \lambda^3} = E \times \frac{1}{V} = \frac{e^2}{2\varepsilon_0 \alpha \lambda} \times \frac{1}{\left(\dfrac{\alpha^5 \lambda^3}{2\pi^2} \right)} \tag{3.16}$$

It is important to note here that this volume in no way represents an actual volume of the related particle. It is by structure the *theoretical stationary isotropic volume* that the incompressible oscillating kinetic energy quantum would occupy if it was immobilized as a sphere of isotropic density. Metaphorically speaking, it amounts to bundling up all of the leaves in a tree into the smallest possible uniformly isotropic sphere to more easily calculate the limit volume and density of the material of which the leaves are made, which allows, in context, determining an electromagnetic particle's energy absolute limit density parameters, beyond which they cannot possibly be increased.

In turn, the definitions of Equations (3.13) allowed defining in Reference ([30], [8] Chapter 4) the electric and magnetic fields corresponding to the invariant rest mass of the electron, separately from those of its carrying energy, by applying the electron Compton wavelength to the fields definitions of Equations (3.13):

$$\mathbf{B} = \frac{\mu_0 \pi e c}{\alpha^3 \lambda_C^2} \qquad \text{and} \qquad \mathbf{E} = \frac{\pi e}{\varepsilon_0 \alpha^3 \lambda_C^2} \tag{3.17}$$

LC Equation (3.6) then allowed upgrading Newton's non-relativistic kinetic energy equation $K = mv^2/2$ to full relativistic status by first converting it to its equivalent electromagnetic form in Reference ([42], [8] Chapter 5). This then allowed correcting it according to the electromagnetic structure of Equation

(3.6) to obtain two new relativistic equations fully derived from electromagnetism, provided as Equations (3.18) below; the first of which allows calculating any possible velocity state of any localized electromagnetic elementary particle from velocity zero for an electron at complete rest to the complete range of relativistic velocities of a massive elementary particle, to velocity c for free moving electromagnetic photons ([42], [8] Chapter 5, Equation (33a)), and the second equation allowing calculation of the velocity of any localized massive elementary particle from the separate wavelengths of the energy of its invariant rest mass and that of its carrying energy ([42], [8] Chapter 5, Equation (49)); this latter more restrictive equation being identical to Equation (55) of the same reference derived from Special Relativity equation $E=\gamma m_o c^2$ in Reference ([30], [8] Chapter 4):

$$v = c\frac{\sqrt{4EK + K^2}}{2E + K} \quad \text{and} \quad v = c\frac{\sqrt{4\lambda\lambda_C + \lambda_C{}^2}}{2\lambda + \lambda_C} \tag{3.18}$$

From the addition of magnetic fields Equations (3.13) and (3.17) for both the rest mass of the electron and its carrying energy, the following equation was then established as Equation (49) in Reference ([30], [8] Chapter 4) to obtain the magnetic field equation for the electron in motion:

$$\mathbf{B} = \frac{\pi \mu_0 e c}{\alpha^3} \frac{\left(\lambda^2 + \lambda_C{}^2\right)}{\lambda^2 \lambda_C{}^2} \tag{3.19}$$

which incidentally exactly corresponds to the magnetic field related to Marmet's Equation ([29], Equation (M-23)) derived from the Biot-Savart equation.

From the product of Equation (3.19), for the electron in motion, by Equation (3.18), for calculating relativistic velocities from wavelengths, electric field Equation (3.20) was then established for the electron in motion in straight line at any velocity. The known relation $\mu_o c^2 = 1/\varepsilon_o$ allowed establishing this equation in a simple manner by substitution, since the product of Equations (3.18) and (3.19) exactly corresponds to the right side of the standard equation for calculating motion in straight line of a charged particle $E=vB$ stemming from the Lorentz equation:

$$\mathbf{E} = \frac{\pi e}{\varepsilon_0 \alpha^3} \frac{\left(\lambda^2 + \lambda_C{}^2\right)}{\lambda^2 \lambda_C{}^2} \frac{\sqrt{\lambda_C\left(4\lambda + \lambda_C\right)}}{\left(2\lambda + \lambda_C\right)} \tag{3.20}$$

Formal establishment of Equation (3.20) actually requires a complex vectorial product in the trispatial geometry, which is described in Reference ([34], [8] Chapter 19) and that remains to be established.

3.16. Mechanical explanation to e⁺e⁻ pair production from the decoupling of 1.022 MeV electromagnetic photons in the trispatial geometry

Analyzing Equation (3.6) with respect to the greatly increased set of orthogonal geometric possibilities allowed by the expanded trispatial geometry also allows establishing a mechanical explanation to the conversion of massless electromagnetic photons to massive electron-positron pairs while preserving the cyclic oscillation of the magnetic aspect of their rest mass energy between increasing spherical presence from zero presence to maximum spherical presence, followed by decreasing spherical presence to zero presence at the electron invariant rest mass energy frequency, which is a critically important feature of the cyclic spin orientation reversal of self-sustaining localized electromagnetic elementary particles' magnetic fields brought to light in the trispatial geometry ([31], [8] Chapter 11), which will be put in perspective further on.

Given that the total complement of energy making up the invariant rest masses of both 0.511 MeV/c² particles generated possess omnidirectional inertia (electromagnetic mass) after conversion of a 1.022 MeV electromagnetic photon, this also means that the natural conversion process allows for the unidirectional half of the photon's energy to mechanically acquire this property of omnidirectional inertia. The manner in which this is accomplished during the conversion process in the trispatial geometry, as well as how the opposite signs of the charges of both particles are acquired, is analyzed in Reference ([31], [8] Chapter 11).

The trispatial LC equation for the electron at rest can be formulated as follows:

$$E\vec{0} = m_e c^2 \vec{0} = \left[\frac{hc}{2\lambda_c}\right]_Y \vec{J}\overleftarrow{i} + \left(\begin{array}{l} 2\left[\dfrac{(e')^2}{4C_C}\right]_X (\,\vec{I}\vec{j},\vec{I}\overleftarrow{j}\,)\cos^2(\omega t) \\ + \left[\dfrac{L_C i_C^{\,2}}{2}\right]_Z \overleftrightarrow{K}\,\sin^2(\omega t) \end{array} \right) \tag{3.21}$$

where λ_c is the electron Compton wavelength. In the trispatial space geometry, the rest mass equation for the positron is identical to Equation (3.21) for the electron, except for a 180° orientation reversal of sub-unit-vector (*i*) in expression (*J-i*) within electrostatic Y-space, which refers to the reversal of the sign of its unit charge with respect to that of the electron ([31], [8] Chapter 11).

3.17. The Coulomb force

Considerations on the possible origin of the momentum related translational kinetic energy that propels elementary charged particles such as electrons lead to observe that at the submicroscopic level, kinetic energy is induced in these particles exclusively as a function of the distance separating them. It is also well verified that the only known force able to induce kinetic energy in free moving charged particles is the well known Coulomb force.

Although established more than 200 years ago by C.A. Coulomb, the exhaustively confirmed Coulomb law which is in action between charged particles as a function of the inverse square of the distance separating them seems to have progressively become invisible in the background of the quantum electrodynamics method (QED), even if the Coulomb equation is an integral part of Maxwell's first equation, that is, Gauss' equation for the electric field, from which it can easily be derived ([32], [8] Chapter 14).

The Coulomb force is indeed a critically important component of every *virtual photon* in QED, but metaphorically cut into so many little pieces that it now attracts little attention. Metaphorically speaking, QED causes us to pay attention to every individual pixel in a metaphorical 4K screen that would represent the submicroscopic level, but if we mentally pull back sufficiently, its infinitesimally progressive action can be observed again.

From observations made at our macroscopic level, the traditional concept of *force* was historically established by Newton as a mutual action between two massive bodies, in the sense that "*when a body exerts a force on a second body, the second body always exerts a force on the first*" ([57], p. 87). Newton established this conclusion as his third law of motion, stating that the mutual actions of two massive bodies on each other are always equal.

Considering each of these bodies separately, the force is then defined as being the interaction that changes the momentum of a body as a function of the time that this interaction is applied to it. This led to defining force as the product of the mass of a body by its acceleration, that is, its changing velocity (*F=ma*); and to define its momentum at any given instant as the product of its mass by its instantaneous velocity ($p=mv$).

This observed *apparent attraction* as a function of the inverse square of the distance between massive bodies that are not in contact with each other, then resulted in *force* being directly related to a natural increase in translational momentum of the body, without any immediate need to refer to the simultaneousness of the increase of its translational kinetic energy as a function of the diminishing distance between the bodies involved, which is obtained by multiplying the force by the distance between the bodies at any given moment, since acceleration is represented by the squared momentary velocity divided by the corresponding instantaneous distance ($a=v^2/r$), which results in the total amount of energy momentarily induced in the body at this specific distance to be ($E=mv^2$), a total amount of induced kinetic energy that Leibnitz considered the real effect of application of a force, as mentioned previously ([57], p. 222), quantity which incidentally is twice the amount associated with the translational momentum (p), which on its part is traditionally calculated by replacing (v) by (p/m) in the classical kinetic energy Equation ($K=mv^2/2$), giving ($K= p^2/2m$) ([8], p. 134).

From the relativistic perspective, the reason for the difference between these two energy measuring methods is that ($E=\gamma m_o v^2$) also includes the induced energy that converts to the velocity related momentary relativistic mass increment that was transversely measured by Walter Kauffman when he deflected relativistically moving electrons in a bubble chamber at the turn of the 20th century [64], and that was established by Paul Marmet as corresponding to the relativistic magnetic mass increment as represented in Equation (3.8), while ($K=\gamma m_o v^2/2$) provides only the correct amount of momentum related translational kinetic energy that sustains the velocity of the total relativistic mass, that is, an amount of unidirectional kinetic energy that turns out by structure to correspond to exactly half of the total amount of kinetic energy that must be induced in the electron in excess of its invariant rest mass energy for it to move at the corresponding relativistic velocity, as analyzed in Reference ([42], [8] Chapter 5), and represented in Equation (3.8).

It must be put in perspective that these definitions, quite useful at our macroscopic level when applied to massive macroscopic bodies, were established before it was discovered that the force in action between charged elementary particles actually induces kinetic energy in these particles due to the fact that they are electrically charged, so in the absence of this information discovered later, the same definitions of force and momentum were applied by default to the Coulomb force as applicable to these elementary massive subcomponents of atoms, without taking into account that besides their mass, they also possess an electrical charge, which is precisely the characteristic related to energy induction in electromagnetism.

The Coulomb force was thus defined in the following manner:

> *"The force of attraction or repulsion between two point charges is directly proportional to the product of the charges and inversely proportional to the square of the distance between them."* ([17], p.462).

But deep analysis of the Coulomb force in light of the internal electromagnetic energy structure of the carrying energy amounts induced in charged particles such as electrons and positrons revealed in the trispatial geometry, and of the variation of these amounts as distances vary between charged particles, reveals that the force itself does not directly attract nor repel in the manner that it is currently defined to operate, but that it only adiabatically induces kinetic energy in electrically charged elementary particles, and that it is the unidirectional momentum related component of this adiabatic kinetic energy that vectorially orients itself to cause charged particles to translationally tend to move toward each other in case of opposite signs charges, or away from each other in case of same sign charges, when the particles are not captive in the various stable electromagnetic resonance equilibrium states allowed in atomic structures, states into which this translational motion is hindered even if the momentum related kinetic energy still remains adiabatically induced, as analyzed in Reference ([45], [8] Chapter 16). This adiabatically maintained presence of kinetic energy will be analyzed further on.

This brings to light that the Coulomb force would not really be a *force of attraction or repulsion* as traditionally defined, but would rather be a *force of adiabatic kinetic energy induction* that would adiabatically and continuously induce kinetic energy in elementary charged particles, whether they are moving or not, which would make this force a *yet-to-be-correctly-understood-active-agent* that would be universally ambient in the background, so to speak, and consequently that it would not need to travel at any velocity to simultaneously act on all existing charged particles in the universe, but would only increase or decrease the amounts of this adiabatically induced kinetic energy in an infinitesimally progressive manner whenever charged particles happen to be in distance varying motion with respect to each other.

Moreover, Marmet's discovery and the observation confirmed by the Kaufmann experiment that half of any carrying energy quantum induced in electrons converts to mass, reveal that not only does the Coulomb force induce the momentum related translational energy of elementary charged particles, it also induces actual mass, made up of the electromagnetically oscillating other half of the induced carrying energy, as represented with Equations (3.8) to (3.10) and as established in References ([42], [8] Chapter 5) ([30], [8] Chapter 4).

From this perspective, and given that this carrying kinetic energy needs to be induced in charged particles *before* any related motion can becomes possible, this means that no motion of the charged particles is required for the Coulomb force to adiabatically induce kinetic energy in them as a function of the distance, and that this energy remains induced even if the related velocity is prevented from being expressed when the particles are captive in stationary orbital resonance states, which are states of induced momentum kinetic energy that the classical concept of momentum, thus also of the Lagrangian and the Hamiltonian, clearly do not account for since its related translational velocity is then forcibly reduced to zero, or averages out to zero for electrons captive in such axial resonance states.

Also, the currently accepted conception is that the Coulomb force would be in action in the hydrogen atom between the electron and the *proton*. This conclusion disregards the fact that the proton is not an elementary charged particle, but a system of elementary charged particles, just like the solar system not a single body, but a system of smaller massive astronomical bodies.

Regrettably, 50 years after that this major discovery was experimentally confirmed at the Stanford linear accelerator in 1968 [21], it seems that few introductory textbooks to particle physics clearly mentions this discovery with proper reference, but instead continue referring to protons and neutrons as being elementary particles, which induces a high level of confusion in the community in this regard.

Obviously, the solar system is a system whose internal structure is defined by planets stabilized on orbits about a central star, and just as obviously since the 1960's, the proton is known to be a system whose internal structure is defined by interacting elementary particles that are charged, massive, scatterable and point-like behaving just like the electron, that were named up quark and down quark, that are electromagnetically stabilized into least action equilibrium resonance states. See also Section 1.23 on this issue.

So since the Coulomb force can act only between electrically charged particles, it obviously can be interacting only between the charged electron and the charged up and down quarks that are captive inside the proton structure. So these 3 particles are the only stable interacting charged and massive elementary particles that can be identified as the physical building blocks of all atoms in the universe, instead of the three that are still often erroneously referred to as being the three fundamental elementary particles set defining the inner structure of atoms: electron, proton and neutron.

Consequently, from the electromagnetic perspective, the hydrogen atom *is not an interacting two-massive-body system* as it still is currently considered, but rather *a four-charged-electromagnetic-particle system* stabilized in least action electromagnetic resonance states.

In light of these considerations, a tentatively more precise definition of the Coulomb force could be formulated in the following manner, for example:

> *"The Coulomb force adiabatically and continuously induces kinetic energy in elementary charged particles as a function of the inverse square of the distance separating them, thus inducing in each charged particle an accompanying energy quantum whose unidirectional half is vectorially oriented so that charged particles tend to close in on each other if they have opposite signs charges, and move away from each other if they have identical sign charges, when not captive in the various resonance states allowed in atoms, and to apply pressure in these vectorial directions when their motion is inhibited by local electromagnetic equilibrium states."*
> ([45], [8] Chapter 16).

So from the submicroscopic perspective, it would then appear that it is not the macroscopic bodies themselves that are subject to a force, but the individual charged and massive point-like behaving electromagnetic elementary particles whose sum of masses makes up the total masses of macroscopic bodies, and that the only force that can act on them would be by structure the so-called *"Coulomb force"*, which would then not be an attractive and repulsive force as initially defined by similarity with the apparent inverse square attraction force between macroscopic masses that was the only possible interpretation in Newton's time, but would rather be an underlying *adiabatic-kinetic-energy-inducing-yet-to-be-correctly-understood-active-agent*, that we name the "*Coulomb force*", which could be by very nature permanently and statically present in the universe and in permanent action between all charged elementary particles in existence.

This means that the kinetic energy induced in any pair of charged particles is inversely proportional to the distance separating them irrespective of the time elapsed, if they are maintained at a fixed distance from each other, and that it adiabatically varies in both particles if they are in motion relative to each other, irrespective of their relative velocity and irrespective of the time elapsed during the corresponding motion sequence.

In the trispatial geometry, both neutral internal charges of electromagnetic photons would logically acquire opposite vectorial signs on the Y-y/Y-z plane, but would both appear neutral along the perpendicularly oriented Y-x axis along which they do not travel, this latter apparently neutral state being the state observable from the perspective given us from within normal X-space in the case of electromagnetic photons.

This tentatively reformulated definition will now allow describing the Coulomb force adiabatic kinetic energy induction process at play within atoms between the charged and massive elementary particles that they are made of.

However, to simplify the description, the traditional terms of *attraction* and *repulsion* will continue to be used often in this text, but always keeping in mind that *attraction* refers to unidirectional carrying energy being vectorially oriented toward an opposite sign particle, and that *repulsion* refers to unidirectional carrying energy being hindered in its translational motion.

3.17.1. The concept of Gravitational waves

The ideas of *the impossibility to demonstrate absolute motion* and to define *relative motion* instead are not the only ideas that Einstein seems to have borrowed from this little article published by Poincaré on June 9, 1905 [76] (See Section 3.4 and Subsection 3.5.1). The concept of the *gravitational wave propagating at the speed of light* that Einstein proposed in 1916 is also proposed in this article, in which Poincaré proposed it under the name of *gravific wave* *("onde gravifique")*:

> *"J'ai été d'abord conduit à supposer que la propagation de la gravitation n'est pas instantanée, mais se fait avec la vitesse de la lumière... Quand nous parlerons donc de la position ou de la vitesse du corps attirant, il s'agira de cette position ou de cette vitesse à l'instant où l'onde gravifique est partie de ce corps; quand nous parlerons de la position ou de la vitesse du corps attiré, il s'agira de cette position ou de cette vitesse à l'instant où ce corps attiré a été atteint par l'onde gravifique émanée de l'autre corps; il est clair que le premier instant est antérieur au second."* ([76], p. 492).

Translation:

> *"I was first led to assume that the propagation of gravitation is not instantaneous, but occurs with the speed of light.... So, when we will speak of the position or velocity of the attracting body, it will be the position or velocity at the instant when the* gravific wave *left that body; when we will speak of the position or velocity of the attracted body, it will be the position or velocity at the instant when that attracted body was reached by the gravific wave emanating from the other body; it is clear that the first instant is prior to the second."*

However, the just carried out analysis of the concept of the so-called *Coulomb force* reveals that the adiabatic ΔK momentum energy that propels one body towards another in the universe is not *transmitted* from one body to the other according to the concept conceived by Poincaré, but is permanently and immediately present by structure in each body and that it locally varies with the inverse of the distance separating them, which means that the instantaneity of interaction between bodies is realized by structure contrary to Poincaré's expectation, as described in Sections 1.26 and 1.27, and that *absolute motion is the actual general law of nature,* as analyzed in Subsection 3.5.1, also contrary to his conclusion also stated in his 1905 Note:

> *" Il semble que cette impossibilité de démontrer le mouvement absolu soit une loi générale de la nature."* ([76], p. 489)

Translation:

"It seems that this impossibility of demonstrating absolute motion is a general law of nature."

But to be fair to Poincaré, it was no more possible for him to conceive the idea that momentum as well as transverse mass energy was adiabatically induced as a *physically existing substance* in all elementary particles, than it was possible for Newton to conceive of the idea that the mass of bodies increases with velocity, since no clue in these directions were available from the more limited extent of knowledge accumulated in their respective eras.

3.18. Adiabatic kinetic energy induction in atomic and nuclear structures

Analysis of the manner in which temperature adiabatically increases with depth inside the Earth mass [62] leads to conclude that this increase can only be related to a progressive compression increase with depth of the electronic orbitals about the nuclei of the atoms making up the mass of the Earth, that would shorten the mean distances between the electrons and the up and down quarks that are the only elementary charged sub-components of the nucleons making up these nuclei, which can only increase the amounts of kinetic energy induced in them by the Coulomb force as a function of the inverse square of these shortened distances.

In turn, this leads to observe that for electrons stabilized into such natural least action states, the kinetic energy can only be induced in an adiabatic manner, since this energy varies progressively as distances vary between these charged particles without any of it being emitted to the environment or being contributed by the environment during this natural compression driven distance variation process ([43], [8] Chapter 2). See also Section 1.27 on this issue.

Since all three elementary massive particles that can be detected via non-destructive scattering within all atoms in existence (electrons, up quarks and down quarks) are charged ([34], [8] Chapter 19), this of course means that adiabatic kinetic energy is permanently induced in all of them, whose quantities are clearly measurable for the stable average axial resonance distances that separates them, and that necessarily correspond to the electronic orbitals for electrons, and to nucleonic orbitals for up and down quarks inside nucleons.

An extensively documented case of such a level of adiabatic carrying energy induction by the Coulomb force is that of the ground state orbital of the hydrogen atom:

$$E = \int_{a_0}^{\infty} \frac{1}{4\pi\varepsilon_o} \frac{e^2}{r^2} \cdot dr\,0 = 0 - \frac{1}{4\pi\varepsilon_o} \frac{e^2}{r_o} = \text{-4.3597438 05 E - 18 J (27.2 eV)} \qquad (3.22)$$

where (r_o) is the Bohr radius, which exactly corresponds to the mean distance separating the electron, stabilized in axial resonance state in the ground state orbital, from the charged up and down quarks captive in the central proton of the hydrogen atom.

The previously established internal electromagnetic structure of the carrying energy of the electron described by Equation (3.14) now reveals that half of this adiabatic energy systematically converts to a mass increment, possessing omnidirectional inertia just like the invariant rest mass of the electron, that increases the momentary electron mass, whether the electron is translationally immobilized in this manner, or freely moving at the velocity corresponding to this amount of carrying energy, as confirmed by Kaufmann's transverse mass measurements of electrons moving at relativistic velocities [64].

This measurable state can now be directly related to the difference between the total amount of carrying energy provided by Equation ($E=\gamma m_o v^2$) stemming from acceleration Equation ($F=\gamma m_o a$), and half this amount calculated with Equation ($K=\gamma m_o v^2/2$), that provides only the translational momentum related kinetic energy that propels the total relativistic mass of the particle.

Since the same adiabatic kinetic energy inducing Coulomb force is structurally at play between the charged up and down quarks inside the proton, the adiabatic mass increments due to their immensely higher levels of carrying energy can only be much more important than in even the most energetic electronic orbitals, given the extremely short distances separating them within nucleons' structures.

Close study of nucleons' structures in the frame of the trispatial geometry in light of the unavoidable presence of this permanently induced adiabatic energy, that contributes to increase the mass of elementary particles as a function of these very short axial distances between the up and down quarks, then led to the establishment of trispatial LC equations for the rest mass energy and for the carrying energy levels of these elementary charged and massive particles making up the internal structure of nucleons that are consistent with observation ([43], [8] Chapter 2) ([32], [8] Chapter 14) ([49], [8] Chapter 9).

These equations reveal that the carrying energy level reached for each up and down quark within the proton structure is about 600 times higher than the energy contained in the rest mass of the electron stabilized in the ground state orbital of the hydrogen atom ([32], [8] Chapter 14).

3.19. The cyclic polarity reversal of elementary particles magnetic fields

The oscillating nature of the magnetic component of elementary particles' invariant rest mass energy and also that of their carrying energy as revealed by LC Equations (3.14) and (3.21), makes obvious that, in the trispatial geometry, the physical presence of this magnetic component can only oscillate between zero presence and maximum spherical presence in space and then back to zero presence at the frequency and to a physical spherical extent related to the amount of energy contained in their quanta.

In the trispatial geometry, a *point-like junction area* or *point-like passage area* is located at the center of each localized electromagnetic quantum, which allows its energy to freely circulate between the three thus interconnected spaces as if between communicating vessels, and to locally stabilize in a state of self-sustaining dynamic equilibrium between the three 3-dimensional orthogonal spaces constituting the trispatial geometric complex within which each electromagnetic energy quantum exists (**Figure 3.1**), which is completely described in References ([15], [8] Chapter 6) ([34], [8] Chapter 19), and that allows the energy of the quantum to be described as one unidirectional half remaining in translational momentum orientation within normal X-space for the photon, while the other half electromagnetically oscillates transversely between two separate orthogonal 3-dimensional spaces that are perpendicular with respect to each other and with respect to normal X-space, one of which is identified as Y-space, allowing manifestation of the properties represented by the electric E field, while the other is identified as Z-space, allowing manifestation of the properties represented by the magnetic B field. This latter half of the particle's energy, being longitudinally inert by structure, consequently displays omnidirectional inertia by definition in normal X-space, that is, *electromagnetic mass*.

In the trispatial geometry, this point-like junction is meant to represent the observable and measurable *point-like behavior* of charged elementary particles such as photons or electrons during scattering encounters between these particles in normal space.

In this space geometry, the energy making up the magnetic component of the rest mass of the electron is by structure in constant internal motion, successively in two opposite spherical orientations, from an initial state of zero presence within magnetostatic Z-space at the beginning of each cycle, after having completely transferred into another space of the complex, this motion of the energy will then consist in two distinct phases, the first being a spherical expansion phase as it omnidirectionally enters Z-space through the point-like junction, until maximum radial expansion has been reached. The second phase will consist in a reverse motion inwards through the trispatial junction as an omnidirectional spherical regression of this energy until it has completely evacuated Z-space. This oscillation process redefines the spin of an elementary electromagnetic particle as becoming a property relative to the state of the expansion and regression cycles of the presence of the magnetic energy of all other electromagnetic elementary particles.

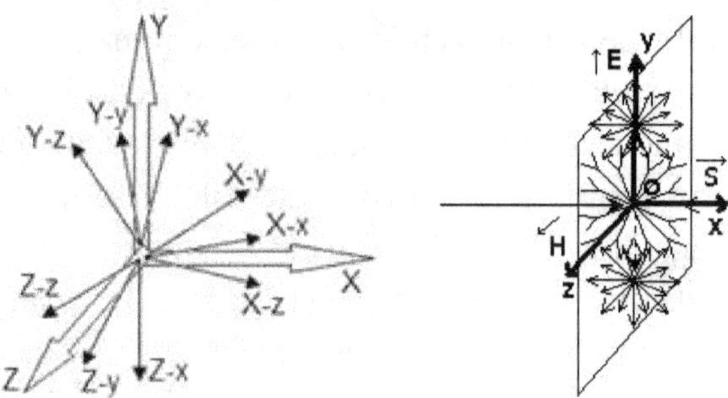

Figure 3.1: The orthogonal structure of the 3-spaces geometry complex, and plane wave reference frame applied to a permanently localized photon.

This behavior also implies that both poles of the magnetic field of an electromagnetic elementary particle have to geometrically coincide by structure with the location of the trispatial point-like junction located at their center. This means that a relative parallel spin alignment between two electrons will occur, for example, when the magnetic presence of the energy of both particles is synchronously in spherical expansion and regression at the same time, which amounts to a spherical inverse cube magnetic repulsion with distance between both magnetic spheres, since the magnetic energies of both particles remain vectorially opposing each other during the whole sequence; while relative antiparallel spin alignment will occur when the magnetic energy presence of one electron is synchronously in its expanding spherical presence phase while that of the other electron is in its regressing spherical presence phase, which amounts to a spherical inverse cube magnetic attraction with distance between both particles, since the energies of both spherical magnetic spheres remain vectorially moving in converging directions during the complete sequence.

Interestingly, the resulting magnetic inverse cube interaction between two electrons forced to interact in parallel repulsive spin orientation was recently experimentally measured by Kotler et al. in 2014 [64].

Moreover, the force that can be calculated between both electrons from the data collected, whose analysis results in the establishment of Equation (3.23), amounts to exactly half the force that can be calculated from the magnetic interaction between two bar magnets within each of which both north and south poles are by structure separated by a measurable distance, and whose force between their simultaneously interacting 2 pairs of poles is calculated with Equation (3.24), which is the standard equation established for dealing with bar magnets ([57],p 93).

$$F = \frac{3\mu_0\mu^2}{4\pi\,d^4} \tag{3.23}$$

$$F = \frac{3\mu_0\mu^2}{2\pi\,d^4} \tag{3.24}$$

This difference in intensity of the force calculated with these two equations appears to directly relate to the fact that within a point-like behaving electromagnetic particle, in which the distance between both poles reduces to zero by structure, both poles can only exist in alternance one at a time in succession, which directly correlates both poles, as well as the relative spin of the particle, to both magnetic energy presence increasing and decreasing phases of the oscillating electromagnetic cycle of the quantum's energy as described by the trispatial LC equations.

Interestingly, the alternating presence of both magnetic poles for magnetic fields for which both poles geometrically coincide like those seemingly observed for the point-like behaving electrons in the Kotler et al. experiment can be quite easily confirmed at our macroscopic level with circular magnets magnetized parallel to thickness, such as loudspeaker magnets ([49], [8] Chapter 9).

Due to the need for the loudspeaker coil to constantly tend to keep perfect perpendicular axial alignment in the central hole of these magnets, this orientation of the magnetic field during the magnetization process forces both poles of the macroscopic magnetic fields that develop about them to geometrically coincide by structure at their geometric center, the proof being that from the data collected from the interaction of such mutually interacting magnets, the force that can be calculated systematically obeys Equation (3.23) just as for the electrons of the Kotler et al. experiment, and cannot be made to obey Equation (3.24) under any circumstance, as analyzed in Reference ([49], [8] Chapter 9), thus demonstrating that during interaction between two such magnets for which both magnetic poles coincide by structure within the magnetic field of each magnet, or in context, between the two point-like behaving electrons of the Kotler et al. experiment, only two poles at a time are simultaneously interacting, and never 4 poles as with bar magnets.

A surprising conclusion of this observed and measured behavior is that magnetic fields inside which both poles geometrically coincide can only be monopolar at any given instant, which means that the magnetic field of the invariant rest mass of electrons as described in the trispatial geometry, and as measured during the Kotler et al. experiment, is a magnetic monopole by structure at any given instant.

Indeed, the Kotler et al. experiment and the circular magnets experiment demonstrate out of any possible doubt that only 2 magnetic poles at a time can simultaneously be interacting during such magnetic encounters between magnetic fields such as those of electrons, that is, only one pole at a time belonging to each particle, which appears to completely validate the cyclic magnetic spin reversal process mandated by the inner structure of electromagnetic particles in the trispatial geometry.

3.19.1 Experimental Proof of Magnetic Poles physical separation within magnetized bars

Recent experiments with the new ferrolens technique developed by Emmanouil Markoulakis [95] reveal a clear physical separation between both poles in a bar magnet.

From the electromagnetic perspective, what seems to happen when the unpaired electrons in the paramagnetic material of the magnetized bar are forced to stabilize in parallel magnetic spin orientation, their individual dynamic magnetic energy that normally oscillating timewise, seems to merge into a common static pool that extends non-spherically in space perpendicularly to the time dimension, due to the extended spatial distribution of the electrons involved, which causes the macroscopic magnetic field thus established to stabilize into a *static spacewise magnetic dipole*, both poles of which are shown by this experiment to remaining physically separated, each of which can be concluded to be a *macroscopic static magnetic monopole*.

3.20. Magnetic fields interaction as a function of identical oscillating frequencies

This state of cyclic magnetic polarity reversal of the magnetic component of the electron brings an entirely new light to the reason why two electrons can associate in antiparallel spin alignment to fill electronic orbitals or to associate in covalent bounding between atoms, despite their electric repulsion as a function of the inverse square of the distance separating them, on account of their identical electrical sign, which at first glance should logically prevent such close association of two electrons.

The answer obviously lies in the fact that their magnetic fields interact as a function of a higher order interaction law than the inverse square electric interaction law **(Figure 3.2)**, which means that when forced by local electromagnetic circumstances to come close enough to each other for the inverse cube magnetic interaction to start overcoming the inverse square interaction, they will easily switch to mutually attractive antiparallel spin alignment, which is a least action state with respect to parallel magnetic spin alignment. This process is analyzed in Reference ([43], [8] Chapter 2).

The same process also explains why a pair of electron and positron that mutually capture in metastable positronium configuration always succeeds in actually spiraling inwards until they meet and convert to electromagnetic photons states as the systematic final stage of the positronium decay process, that benefits from the additional favorable circumstance that contrary to a pair of mutually interacting electrons, both particles also electrically attract as a function of the inverse square law, which easily brings them to the equilibrium point at which the inverse cube magnetic interaction will dominate ([43], [8] Chapter 2). Their respective amounts of carrying energy being equal by structure in the positronium system, their contributing magnetic fields will also oscillate at the same mutual frequency and will not hinder the process in any way.

The success of the antiparallel spin coupling of electron pairs in covalent bounding and electronic orbital pair filling, as well as the final stage of the positronium decay process resulting in the electron and the positron physically joining to convert to electromagnetic photons state is intimately linked in the trispatial geometry to the fact that the magnetic oscillating frequencies of both particles are identical, which allows them to easily fall into perfectly synchronized least action antiparallel magnetic oscillation.

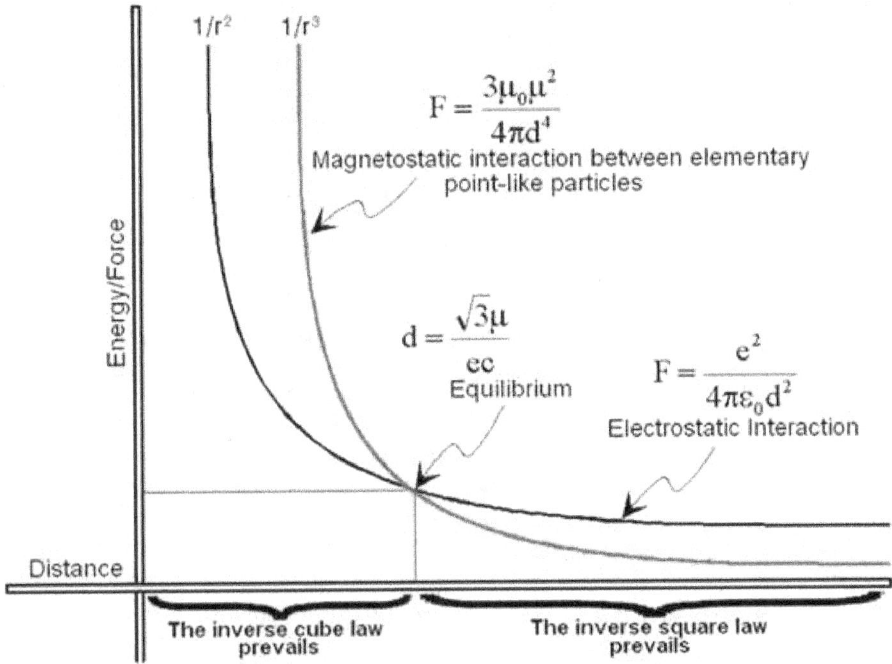

Figure 3.2: Intersecting inverse square and inverse cube interaction curves.

3.21. *Magnetic fields interaction as a function of different oscillating frequencies*

The situation is quite different however when an electron and a proton are interacting to form a hydrogen atom, even though they display equal intensity opposite charges signs similar to those of an electron-positron pair meta-stabilizing into positronium configuration.

The difference lies with the *apparent* unit positive charge of the proton system, which, let us recall, is a system of elementary particles electrically charged, and is not itself a charged particle. The peculiarity with the proton apparent unit charge, which is often overlooked, is that its assumed unit charge is the results of the addition of the fractional charges of its three elementary components, that is, +2/3 +2/3 -1/3 = +1. So this means that the electron is not really electromagnetically interacting with the proton as such, but rather with its three electromagnetic elementary charged inner subcomponents (uud).

Contrary to the positronium case, where both particles' magnetic energy oscillates at the exact same frequency in the trispatial geometry, the hydrogen atom involves the frequencies of the two oscillating magnetic fields of the electron and of its carrying energy on one hand, which are now interacting with the much higher oscillating frequencies of the magnetic fields of the charged inner sub-components of the proton and of their immensely more energetic carrying energy on the other hand ([43], [8] Chapter 2) ([32], [8] Chapter 14). See also Section 2.20.

In the best of cases, the magnetic polarity reversal of the magnetic presence of the most energetic proton inner components occurs more than 600 times during each magnetic presence cycle of the electron magnetic energy (**Figure 3.3**), which, due to the fact that the intensity of the inverse cube magnetic interaction force drops rapidly with increasing distance, results in the magnetic interaction between the electron and the proton inner components becoming predominantly repulsive each time the electron comes closer to the proton than the mean ground orbital distance, which happens to correspond to the distance at which the intensities of both the electric force and the magnetic interaction fall into equilibrium (**Figure 3.2**).

The constant interplay due to the different frequencies of the various magnetic fields involved in inverse cube interaction that oppose the unidirectional momentum energy of the electron that constantly tends to cause the electron to move toward the proton can then only result in the establishment of a stable axial resonance state (**Figure 3.3**) that certainly can be related to Louis de Broglie's initial intuition that electronic orbitals have to be such resonance states, which in the trispatial geometry correspond to the various least action electromagnetic equilibrium states into which elementary charged particles become captive within atomic and nucleonic structures ([43], [8] Chapter 2).

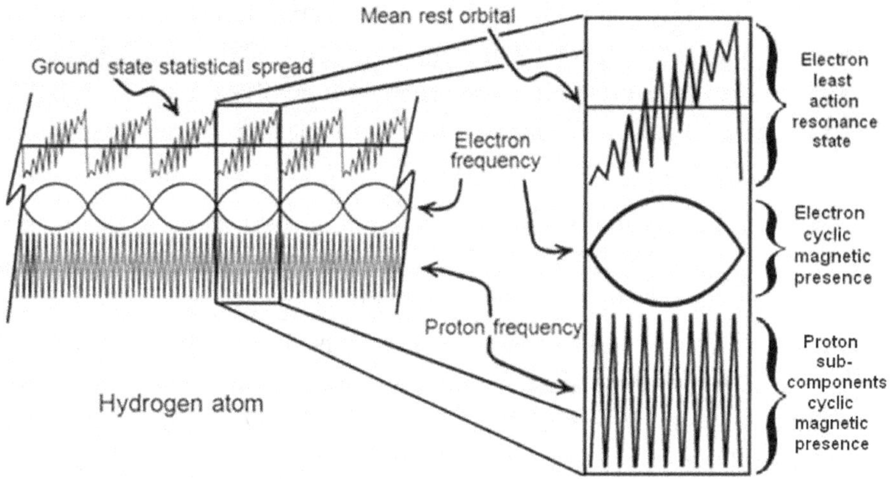

Figure 3.3: Establishment of the least action resonance state of the electron in the hydrogen atom.

The detailed foundation of the mechanics of this electromagnetic resonance state is analyzed in References ([43], [8] Chapter 2) ([32], [8] Chapter 14), and can be summarized as follows. Considering Figure 3.3, the central sequence represents an arbitrary sample of 6 occurrences of the intensity variation of the spherical presence of the electron magnetic energy as a function of its frequency. In a simplified manner, each of these 6 occurrences is symbolically confronted in the lower sequence by the more than 600 occurrences of the spherical intensity variation of the presence of the magnetic energy of only one of the carrying energy quanta of the up and down quarks of the proton as a function of its own frequency. See also Section 2.20.

Figure 3.3 represents the fact that while the electron reverses its spin polarity once, this inner component of the proton reverses its own spin polarity more than 600 times, which means that during each electron spherical magnetic energy presence cycle, the magnetic field of this proton inner component will alternate more than 600 times between being in relative parallel spin alignment with respect to the electron's magnetic field spin orientation, thus repelling it, and being in relative antiparallel spin alignment, thus attracting it.

The least action orbital equilibrium state is consequently established by the fact that the permanently induced unidirectional translational momentum component of the adiabatic carrying energy of the electron, that constantly tends to propel the electron toward the proton, is alternately hindered in its forward motion each time the magnetic interaction function of the inverse cube law becomes repulsive, causing both magnetic spheres involved to repel each other, and is then freed from this counter pressure while the magnetic interaction becomes attractive.

As represented with **Figure 3.3**, during each of the 600 magnetic cycles of the proton inner component, the electron will be axially repelled away from the proton by distance $+d$ during half of the proton inner component magnetic presence cycle during which the spin alignment is parallel, and since the electron will be farther away from the proton as the relation becomes antiparallel for the same duration, there will be a physical impossibility for it to be axially brought back all the way to distance $-d$, given that the inverse cube force will be weaker at this farther location from the proton at the beginning of the antiparallel phase.

Therefore, by structure, due to the more weakly acting inverse cube attraction at the beginning of attractive phase, the electron can be axially brought back only to distance $-(d-\Delta d)$, which will cause it to progressively move away from the proton at each polarity reversal sequence until its own magnetic energy presence falls to zero, moment at which only the electron adiabatic carrying energy translational momentum energy will be at play, causing the electron to move as close to the proton as the inverse square law will bring it until its next magnetic presence cycle initiates and that the whole predominantly repulsive magnetic sequence is initiated again, as represented with **Figure 3.3**.

Of course the actual least action orbital resonance state of the electron in the hydrogen atom or in any other atom will be much more complex than hinted at with this limited example, which is only meant to describe the fundamental mechanics of the process, and will mandatorily involve all such electromagnetic interactions between the magnetic field of the electron and those of all other electromagnetic components captive into nearby atomic and nuclear structures.

Given the mean equilibrium distance that this process causes the electron to stabilize at in the hydrogen atom, it also becomes obvious that the probability distribution of all of the possible instantaneous locations that the electron will stochastically visit about this mean axial distance will be similar to Heisenberg's statistical distribution and will be restricted within axial limits consistent with the fact that the actual amplitude of the volume that the electron can thus visit is dependent on its varying relativistic mass related inertia at any given instant ([43], [8] Chapter 2) ([32], [8] Chapter 14):

$$\int_{-d}^{+d} |\psi|^2 \, dxdydz = 1 \tag{3.25}$$

It seems also entirely reasonable to conclude that the elementary charged up and down quarks making up the scatterable inner structure of protons and neutrons and their carrying energy, which are the only constituting subcomponents of all atomic nuclei in the trispatial geometry, as analyzed in Reference ([32], [8] Chapter 14), would be subject to similar resonance states within their own local least action electromagnetic equilibrium states, that could then also potentially be described by the various methods of quantum mechanics.

3.22. *Resonance states in quantum mechanics and electromagnetism*

It can be observed that quantum mechanics and electromagnetism deal with resonance states from entirely different perspectives, the first at the general level by means of the wave function, that establishes resonance volumes, as for example in a simpler manner in classical mechanics to calculate the volume of space visited by a vibrating guitar string; and the second more directly from the reciprocating mutual induction of electric and magnetic fields as embodied with LRC resonance for example. This is why it appears entirely logical that the inner LC electromagnetic structures that the trispatial geometry allows associating with electromagnetic elementary particles, that allows associating a permanent localization of the moving electron by relating the internal point-like junction of the LC behavior of their energy quantum to their point-like behavior in all scattering encounters, could thus allow the description of the actual electron axial resonance trajectory mechanics within the volumes defined at the general level by the wave function, which, when properly mathematized, could thus provide a fourth quantum mechanics representation that will reconcile permanent localization of the electron with the wave function.

Other studies can also be located that endeavor to directly correlate QM and electromagnetism from the resonance perspective. One example is this interesting study by V.A. Golovko [68] regarding resonance interactions between QM stationary states and electromagnetic wave emission and absorption.

90 years after the identification by Louis de Broglie that electronic orbitals have to be resonance states [4], research on resonance states seems to be resuming in new directions. Another example is this fascinating study on resonance states in the solar corona by Antony Soosaleon involving electric and magnetic fields interaction [96], which proposes a solution to the currently unexplained extreme heat in the solar corona, which is different from that which naturally stems from the trispatial geometry as proposed in Reference ([61], [8] Chapter 15).

3.23. Momentum, the Hamiltonian and the Lagrangian

As mentioned previously, an analysis from the electromagnetic perspective reveals that the progressive adiabatic heat increase with increasing depth in the Earth mass can be related only to an adiabatic compression gradient of the electronic orbitals toward the nuclei of the atoms making up the mass of the Earth as depth increases ([43], [8] Chapter 2). This process mandatorily involves an increase in adiabatic kinetic energy induced by the Coulomb force in all electrons stabilized in the various orbitals, due to the related shortening of the inner atomic axial distances separating them from the nuclei of the atoms to which they belong.

Going back to the origin of the concept of momentum, it can be observed that the concept was intimately tied to motion before the existence of electrically charged and massive elementary particles was discovered and the Coulomb force identified as being the ultimate cause of kinetic energy induction in them, as previously put in perspective.

Although adiabatic processes were already being studied at the time, the idea that momentum related translational kinetic energy could remain induced in bodies stabilized into natural least action dynamic equilibrium states might be related to such adiabatic process obviously did not attract attention, such as the stabilized momentum energy of the massive elementary particles making up the mass of the Earth on its orbit about the Sun.

The initial concept of the presence of kinetic energy as being dependent on motion was then not re-visited and was integrated unchanged into the representations by means of the Lagrangian and then of the Hamiltonian to be applied at the submicroscopic level, even after incorporation of the concept of electromagnetic fields, which perpetuated the assumption that translational motion needs to occur before kinetic energy could even exist and for the related magnetic and electric fields to emerge at the submicroscopic level, instead of concluding that kinetic energy mandatorily had to initially adiabatically exist before motion and the related electric and magnetic fields could emerge from its presence.

This led to the still current perception that momentum related kinetic energy has to convert to *potential energy* so that the process can be seen as conservative, when the motion of electrons is hindered when captured in into atomic structures, which disregards the fact that in physical reality, this momentum related kinetic energy remains adiabatically induced in these electrons even when their motion is inhibited.

It seems that in reality, this adiabatically maintained but translationally hindered momentum kinetic energy continues to *fight* against this hindrance, a constant fight that manifests itself as a permanently maintained axial *pressure* in the vectorial direction of the nucleus against the *counter-pressure* of the predominantly repulsive magnetic interaction between the magnetic energy of the electrons and that of the inner components of the nucleons of which atomic nuclei are made.

Consequently, contrary to current the momentum conservation concept expectations, it would seem that momentum related kinetic energy would be a really physically existing *substance* and that it would behave accordingly. This means that it would not *miraculously* morph into becoming some form of inactive nondescript characteristicsless potential energy when the motion that it sustains is hindered, to just as *miraculously* morph back into becoming active unidirectional kinetic energy when its motion is unhindered as currently represented, but would rather remain constantly present and active even when its motion is hindered, but in a manner that the current concept of momentum / Lagrangian / Hamiltonian is unable to account for.

The consequence of this concept of conservative momentum having been carried on into the Lagrangian and the Hamiltonian without being adapted to account for this fact, is that with respect to the relations between force, motion and matter, classical and relativistic mechanics (CM and RM) keep on dealing with this *real kinetic energy* almost as an afterthought, due to the fact that in CM and RM the only parameter determining momentum besides mass is velocity. Since mass is defined as remaining constant in CM and RM, this makes kinetic energy appear as being an emergent quantity that depends on the prior presence of velocity, and not as a pre-existing primordial quantity that can cause velocity when its motion is not hindered by local electromagnetic circumstances.

In reality, the adiabatic nature of the kinetic energy induced in charged particles rather mandates that in reality, velocity, pressure, charge and mass, can only be emergent properties due to the adiabatically maintained presence of this kinetic energy. Of these four properties of kinetic energy, pressure and the sign of charges are related in the trispatial geometry to the forced inhibition of the translational velocity of the momentum related unidirectional kinetic energy half quantum of elementary electromagnetic particles and of their carrying energy, forcing this unidirectional kinetic energy into configurations that induce these properties; while mass, more precisely defined as being *omnidirectional inertia*, is related to the fact that the transverse electromagnetically oscillating half quantum of any photon or elementary particle carrying energy, and the whole quantum of massive elementary particles, are translationally inert in normal space ([34], [8] Chapter 19)].

It may well be the fact that the kinetic energy of a body is considered to fall to zero when this body is translationally immobilized, in the Hamiltonian / Lagrangian traditional conservative conception of momentum, that made it difficult up to now to clearly identify the nature of these three last properties of kinetic energy, since they seem to be linked to this adiabatically maintained presence, accompanying the invariant rest mass of all massive and charged elementary particles making up all macroscopic massive bodies, of quantities of unreleasable kinetic energy not subject to the Principle of conservation of energy ([43], [8] Chapter 2), and of which the Hamiltonian and Lagrangian, as currently defined, are unable to account for when translational velocity falls to zero, or averages out to zero during such axial resonance states of motion.

3.24. The Submicroscopic Momentum disconnect

It can also be observed that there is a major difference between the definition of momentum applied to classical and relativistic mechanics on one hand, and that applied to electromagnetism, QED and QM on the other hand. This difference relates to the fact that the first two were developed to deal with physical processes at the macroscopic level without taking the electromagnetic properties of elementary particles into account, while the second group was developed to deal with the physical processes at the submicroscopic level of physical reality where there is no choice but to take these properties into account, despite some overlap of both levels by relativistic mechanics and electromagnetism.

What characterizes the first group is that it deals strictly with masses and their observed interactions, mainly at the macroscopic level, without taking into account that their measurable mass at the macroscopic level is only the result of the addition of the individual invariant masses of the charged elementary particles of which they are made and of the massive components of their carrying energy that physically exist at the submicroscopic level. The second group on its part directly deals with the electrically charged electromagnetic elementary particles and their carrying energy without taking into account that the electromagnetic energy that they are made of can exist only in the form of localized self-sustaining quanta involving mutual electric and magnetic induction, which is the fundamental requirement for electromagnetic energy to even exist in electromagnetic theory.

At this submicroscopic level, it was clearly established that the only way for an electron, charged and massive, to be translationally stopped in nature with respect to its environment is for it to be captured by an atom into one of the electromagnetic resonance states that are permitted in this atom; which involves, besides losing its accumulated translational kinetic energy half-quantum as an escaping Bremsstrahlung electromagnetic photon, which is subject to the Principle of conservation of energy, the simultaneous adiabatic induction of the exact same amount of replacement translational momentum kinetic energy, that should also be related to the concept of momentum / Lagrangian / Hamiltonian, mandated by the Coulomb force at this distance from the nucleus, as put in perspective in Reference ([43], [8] Chapter 2), that immediately and synchronously replaces the emitted energy, even if its now hindered translational velocity toward the nucleus now averages out to zero, which is an amount of translationally hindered kinetic energy that will nevertheless remain induced in the particle for as long as the particle will remain in this related least action resonance state.

In all such cases, instead of converting to inactive virtual *potential energy* as currently assumed with the traditional concept of momentum / Hamiltonian / Lagrangian, when the translational velocity of elementary charged particles is hindered, the induced kinetic energy can only remain active, applying continuous *pressure* in the same vectorial direction.

The consequence of the current definition of momentum as being conservative is that in all domains of conventional physics, that is, classical and relativistic mechanics, electromagnetism, electrodynamics and quantum physics, kinetic energy is deemed to exist only if translational motion occurs for a mass at the macroscopic level and for a charged and massive elementary particle at the submicroscopic level, and is viewed by structure as non-existent when translational velocity is reduced to zero, which is where there is such an irreconcilable disconnect between the traditional concept of momentum / Hamiltonian / Lagrangian and the real state of adiabatic kinetic energy induction in all electrically charged elementary particles captive in least action electromagnetic equilibrium states at the submicroscopic level.

3.25. *Diabatic and Adiabatic processes*

Few studies have been carried out regarding adiabatic processes at the submicroscopic level that could be related to the Hamiltonian, and all of them involve changes of state due to changes in ambient conditions as a function of time. These time based changes are covered by the adiabatic theorem that was established by Max Born and Vladimir Fock in 1928 [97]. It is to be noted that these conclusions have not been re-visited since, and that no traceable study seems to have been carried out after the confirmed discovery that nucleons are not elementary particles, but are complex systems made of charged and massive elementary particles also stabilized into least action electromagnetic equilibrium states exactly like electrons in their orbital states.

The Born-Fock analysis concluded that rapid changes in ambient conditions (varying ambient magnetic fields, for example) prevent systems from adapting their configurations, which causes them to remain unchanged, processes that they termed "*diabatic processes*", leaving the final Hamiltonian in a state equivalent to its initial state.

Alternately, they concluded that gradual changes of ambient conditions allow systems to adapt their configurations, which results in their probability densities to be modified during these processes, termed "*adiabatic processes*", causing their final Hamiltonian to stabilize in a state different from their initial Hamiltonian.

Close comparison of these conclusions with the conclusions arrived at in References ([43], [8] Chapter 2) ([49], [8] Chapter 9), in the case of the stability of the hydrogen ground state orbital, reveals that the systems that they were referring to are the least action resonance volumes whose shapes and amplitudes can be determined by the wave function, each of which corresponding to one of the stable least action electromagnetic resonance states into which electrons become captive in atoms.

The related conclusion, drawn in Reference ([43], [8] Chapter 2), is that the wave function describes the shape of the volumes occupied by the statistical spread of the positions that an electron can possibly occupy in the various orbital configurations as a function of local circumstances, as defined originally, while the resonance mechanics previously described explains the existence of these volumes and their elaboration as a function of time, as localized electrons are forced into constant resonance axial motion in reaction to the local magnetic interaction fluctuations; their permanent localization during the resonance process being established by correlating their point-like physical presence in space with the point-like trispatial junction located in their center in the trispatial geometry. See Section 2.20.

Consequently, it can be observed that the Hamiltonian as currently defined deals at the general level with how the volume occupied by the statistical spread of one state can be made to evolve into one of the other authorized volumes, but in no way deals with the continued presence of the translationally hindered adiabatically induced unidirectional half of the electron carrying kinetic energy, which is now mostly acting axially toward the nucleus, while captive on a clearly definable axial resonance trajectory about a mean distance from the nucleus, while alternating between diminished and increased adiabatic induced energy intensity sequences, as the electron is forcibly pushed away from and then released to move back toward the nucleus within the volumes determined by the wave function ([43], [8] Chapter 2).

3.26. Repairing the Submicroscopic Momentum Disconnect

It is clear from the analysis carried out in References ([43], [8] Chapter 2) ([52], [8] Chapter 10) that the transversely oscillating electromagnetic half of the induced adiabatic carrying energy of charged elementary particles, that provides to the particle its omnidirectionally inert related mass increment, is not affected whether or not its unidirectional other half is prevented from being expressed as a translational velocity of the particle, while it is stabilized into one of the possible orbital resonance states in atoms.

On its part, the natural motion of the unidirectional half of the induced energy can be resisted translationally by local electromagnetic equilibrium states in a way that can only lead to the hindered velocity being expressed as a replacing *pressure* constantly exerted in the direction of the nucleus, given the opposite signs of the charges of the electron and that of the sum of the charges of the nuclei's nucleon internal charged components, which determine the vectorial direction of application of this *pressure*.

The predominantly repulsive magnetic interaction that counters the motion of the electron toward the nucleus can logically only be by nature a *contact* resistance between the spherically oscillating kinetic energy magnetic spheres of the particles involved, and of their carrying energy, *bumping* against each other, so to speak, within magnetostatic Z-space ([15], [8] Chapter 6), which provides an elastic contact surface that opposes by structure, to the electron, the same type of hindrance to moving closer to the center of mass of the atom that the surface of the Earth opposes to bodies lying on the ground to moving closer to the center of mass of the Earth.

From the strict electromagnetic perspective, it must always be kept in mind that all macroscopic bodies lying on the ground, as well as all of the matter of which the ground is made at the surface of the Earth, are ultimately made up of atoms, whose ultimate building blocks are only electrons, up quarks and down quarks, which are the only stable scatterable point-like behaving, electrically charged and massive electromagnetic elementary particles that ever were detected inside atomic and nuclear structures by means of non-destructive scattering, and that are the only components of matter that can be induced with kinetic energy by the Coulomb force.

The charged particles making up these bodies lying at the surface of the Earth are consequently also in constant Coulomb force inverse square interaction function of distance with the charged particles making up the mass of the Earth, which consequently find themselves in the same situation as an electron being attracted to a proton by the Coulomb force in a hydrogen atom, even while being captive of each other in various least action electromagnetic equilibrium states to form these macroscopic masses ([43], [8] Chapter 2).

In other words, this *pressure*, now replacing the electron's inhibited velocity, in the direction of application of the unidirectional energy of its carrier-photon toward the proton, amounts to a *gravitational force* in newtons (N) that the electron is applying toward the nucleus while captive at mean ground state orbital distance from the hydrogen atom.

In this regard, Reference ([44], [8] Chapter 7) clearly establishes the mutual identity of all classical force equations by mathematically demonstrating that they all can be converted to *F=ma*, which includes the establishment of the identity between the macroscopic gravitational force with the Coulomb force, after having clarified that the gravitational constant that must be used in any natural axially structured many-bodies system must take into account the orbital parameters specific to the relative order of magnitude of that system for it to remain coherent with observed reality, whence the establishment of a gravitational constant specific to the hydrogen atom, ref: ([44], [8] Chapter 7, Equation (13)) reproduced here for convenience:

$$G_p = \frac{4\pi^2 r_o^3}{M_p T^2} = 1.51417298\,3\text{E}29 \text{ N} \bullet \text{m}^2/\text{kg}^2 \tag{3.26}$$

where M_p= 1.67262158E-27 kg is the mass of the proton, r_o=5.291772083E-11 m is the mean hydrogen ground state orbital radius, and T= 1.519829851E-16 s is the time that would be required for the electron to orbit the proton one time at distance (r_o) if it could so translate; in replacement of (M), the mass of the Sun, (r), the mean distance between the Earth and the Sun, and (T), the time taken for one orbit of the Earth about the Sun, which are the values embedded into the standard definition of astronomical constant G ([44], [8] Chapter 7).

What allows using the potential time that the electron would take to travel once about the proton at distance (r_o) from the proton, as theoretically proposed in the Bohr atom, is the fact that the correct level of energy that would allow the electron to really move at the corresponding velocity is permanently induced by the Coulomb force at this distance of the nucleus of the hydrogen atom. So this time element is coherent with the quantity of motion of the fully expressed corresponding momentum even with its current definition, and can be calculated from the frequency of the carrying energy adiabatically induced at the mean Bohr radius (4.359743808E-18 j), which corresponds to the number of times the electron would orbit the nucleus at distance (r_o) in 1 second at the corresponding velocity:

$$T = 1 \text{ sec} / 6.57968391E15 \text{ Hz} = 1.519829851E\text{-}16 \text{ sec.} \tag{3.27}$$

This velocity replacing *pressure* now oriented toward the nucleus corresponds to the well known *force* of 8.238721759E-8 newtons applicable to the mean hydrogen ground state orbital, and is put in correct perspective as calculated in Reference ([44], [8] Chapter 7, Equation (14)), reproduced here for convenience:

$$F_g = \frac{e^2}{4\pi\varepsilon_o r_o^{\,2}} = G_p \frac{M_p m_e}{r_o^{\,2}} = 8.238721759E\text{-}8N \tag{3.28}$$

3.27. Conclusion

Observing that physical reality was of necessity explored from our macroscopic perceptions digging inwards toward the submicroscopic level as more and more understanding was gained about the nature of matter, mass and energy, which eventually led to important issues remaining unresolved despite our current rather deep knowledge base, it appeared interesting to attempt addressing these issues from what was now known about the submicroscopic level, building upwards toward our macroscopic level.

Analysis of this knowledge base then allowed identifying the electromagnetic properties of energy as ruling this ultimate bottom of the submicroscopic level of physical reality, where only one energy inducing force can be identified, which is the Coulomb force as previously put in perspective.

This perspective also brings to light two major aspects of electromagnetic elementary particles that turn out not to have yet been taken account of in the currently useful theories that were developed over time, which is the fact that the current mechanics theories do not take into account the physical presence of the elementary charged and massive particles of which they are made and of the consequences of their individual motion on the state of motion of the macroscopic bodies to which they belong, as exemplified by the issue that this situation raises with respect to macroscopic rotating bodies for example, and the fact that quantum mechanics and electromagnetism do not yet integrate the mandatory internal mutual induction of the electric and magnetic aspects of electromagnetic energy quanta in a manner that mechanically explains why these quanta can be self-sustaining in a localized manner and behave point-like during scattering encounters.

Interestingly, this proposed alternate foundation of physical reality seems to directly correlate with the zero-point energy level of the quantum vacuum concept that postulates a hypothetical uniform zero point energy excitation level of the quantum vacuum at the beginning of the universe, which is the foundation QFT. The main difference is that this alternate foundation proposes a hypothetical uniform zero energy level in space at the beginning of the universe, that then provides a continuous infinitesimally progressive interaction alternative that offers seamlessly workable mechanical solutions that QFT does not provide, which are, among other benefits, a Maxwell equations compliant mechanical description of the internal self-sustaining mutual induction of the electric and magnetic fields of the localized energy quantum constituting each electromagnetic photon ([15], [8] Chapter 6) and of the invariant rest mass of each charged and massive elementary particle ([31], [8] Chapter 11) ([32], [8] Chapter 14), clear separation of the also electromagnetic carrying energy of elementary particles from the energy making up their invariant rest mass ([42], [8] Chapter 5) ([30], [8] Chapter 4), which allows becoming aware of the adiabatic nature of this carrying energy induced in all charged elementary particles as a function of the distances separating them ([43], [8] Chapter 2), and an electromagnetism compliant mechanical explanation of the stability of electronic as well as nucleonic resonance orbitals states ([43], [8] Chapter 2) ([49], [8] Chapter 9). See also Chapter 2.

Considering that at the beginning of the universe, the ultimate bottom of the submicroscopic level would have been an energyless static empty vacuum devoid of any charged particles that the Coulomb force could have caused to interact, instead of the quantum vacuum zero energy point proposed by QFT that creates particles-antiparticles pairs by means of assumed spontaneous natural quantum vacuum fluctuations, obviously raises the question of how the first electromagnetic photons could have appeared at the origin of the universe, when no charged particles even existed to be accelerated to eventually liberate the first Bremsstrahlung photons that are required from this perspective to mutually destabilize in a process whose existence was confirmed by K. McDonald et al. in 1997 at the SLAC facility [24] into producing the first ever electron-positron pairs that could then be accelerated by this inducing force and be induced with the first ever adiabatic carrying energy quanta, eventually leading to the production of the first nucleons and first hydrogen atoms.

This issue, that of course remains pending, is analyzed in Reference ([56], [8] Chapter 17) where it is tentatively addressed by the idea that the constancy of the flow of time may also be kinetic energy driven and that some punctual event in the far past could have momentarily impeded its motion, thus triggering the release in space of the initial electromagnetic quanta as energetic Bremsstrahlung photons, thus initiating a charged particles generation process that would still be ongoing ([45], [8] Chapter 16) ([61], [8] Chapter 15).

The concept of electromagnetic elementary particles self-energy of QFT is replaced by the mechanically definable concept of self-sustaining mutual induction of elementary particles' electric and magnetic aspects of the energy of the localized quanta of elementary charged particles ([15], [8] Chapter 6) ([32], [8] Chapter 14) ([31], [8] Chapter 11).

This force being statically present and in permanent action between each pair of charged particles, each occurrence of such interaction between charged pairs can then be seen as one unit occurrence among the multitude of such occurrences constituting a universal gradient made strictly of the addition of all such active occurrences between all existing charge pairs in the universe. Contrary to QFT, where the presence of individual excited states affects the intensity of the local energy gradient, the presence of two electromagnetic particles is required for each discrete unit Coulomb force interaction occurrence to exist in the universal gradient, so the gradient is one of intensity of these interaction occurrences and not directly one of energy intensities, or density, as in QFT.

Although the gradient involves the Coulomb force, it does not involve the traditional continuous electric field associated with this force, but uniquely the limited set of all really existing discrete interaction occurrences at play between the really existing charges in the universe as a discontinuous assembly of individual occurrences.

It becomes possible now to separate this gradient into four ranges of intensity levels, whose limits correspond to the various resonance intensity ranges that can be identified in nature. As put in perspective in Reference ([45], [8] Chapter 16), the most intense level is determined by the resonance states characterizing charged elementary particles interactions within nucleons. The second level applies to nucleons stabilization within nuclei. The third level applies to electronic resonance states within atoms and molecules, as well as between atoms and molecules in direct contact with each other in any local accumulations of matter. And finally, a forth and ultimate level of intensity applies to all atoms, molecules and larger bodies in state of freefall, a category that includes macroscopic orbits stabilization at the astronomical level.

These various ranges of intensity of induction of adiabatic carrying energy by the Coulomb force, one of whose major component is the permanently induced adiabatic mass increment that it provides for each existing charged particle, can then be directly related to the 4 forces of the Standard model as put in perspective in Reference ([45], [8] Chapter 16), four forces which then turn out to only be approximate alternate representations of the various intensity ranges of application of the same underlying Coulomb adiabatic energy inducing force.

It is consequently at this point that a clear relationship can be established between quantum mechanics and this global gravitational gradient since the wave function establishes with precision the locations and shapes of the volumes within which each electron stabilizes into its least action electromagnetic resonance equilibrium orbital states by means of one interaction occurrence of the gradient, as clarified in Reference ([43], [8] Chapter 2) and is consequently related to the third intensity range of the universal interactions intensity gradient. This interaction occurrence can then be recognized as a one local occurrence of the classical *gravitational force* acting as a function of the inverse square of the distance between the electron and each of the charged elementary subcomponents of the nucleus, each of them corresponding to an occurrence of the tertiary attractors category described in Reference ([45], [8] Chapter 16).

Each element of the global gradient contributes to the distance coupled adiabatic energy induction variations imposed on the charged particles by the local dynamic circumstances that define their local effective masses. That is, dynamic circumstances that evolve over time according to the rate of matter accumulation in stellar bodies, one of the most remarkable process of which, is the mechanical process that relates the stars ignition threshold to the progressive adiabatic compression of the hydrogen atoms ground state orbitals as depth increases toward the center of protostar masses, due to accumulation of primordial hydrogen atoms, up to the point at which their ground state orbital reaches the axial distance within hydrogen atoms at the center of such masses, that provides them with the energy level that triggers the neutron nucleogenesis process that initiates the fusion process, as also analyzed in Reference ([45], [8] Chapter 16).

Interestingly, given that up and down quarks resonance states inside nucleons are by structure submitted to the same electromagnetic resonance mechanics as electrons in atomic orbitals, it can be concluded that the various wave function representations of quantum mechanics could be adapted to directly apply to them within nucleons in a much more integrated and satisfactory way than QCD allows, which would associate quantum mechanics to the most intense intensity level of the gravitational gradient.

Finally, given the variability function of distance of the size of the adiabatic mass increments which are part of any amount of carrying energy induced in charged elementary particles by the Coulomb force, represented in Equations (3.10) and (3.14) as determined from the analyses carried out in References ([42], [8] Chapter 5) ([30], [8] Chapter 4), it can be observed that the sum of the experimentally confirmed maximum invariant rest masses of the three up and down quarks constituting the interacting inner structure of protons and neutrons, amounts to barely from 2 to 2.4% of the measured masses of these nucleons, and that consequently more than 97% of the masses of all massive bodies in existence can only be of adiabatic origin and are thus part of the carrier-photons of charged and massive elementary electromagnetic particles ([43], [8] Chapter 2) ([32], [8] Chapter 14).

This means that the mass of nucleons can vary as a function of the local intensity of the gravitational gradient and that more than 97% of the measurable mass in the universe is adiabatically induced by the Coulomb force in this manner, which reveals that the mass of astronomical bodies is also variable as a function of the distances separating them ([45], [8] Chapter 16).

Appendix A

A.1. Derivation of the Relativistic Energy-Momentum Equation

Reference [17] mentions on page 835 that combining equations $E=\gamma m_o c^2$ and $p=\gamma m_o v$ should generate the complete relativistic energy-momentum Equation (2.41), but it does not offer the detailed derivation of this equation: $E^2=(pc)^2+(mc^2)^2$.

So, here is for convenience the complete step by step derivation of this famous equation:

$$E = \gamma m_0 c^2 \qquad\qquad p = \gamma m_0 v \qquad (A.0)$$

$$\frac{E}{m_0 c^2} = \frac{1}{\sqrt{1-v^2/c^2}} \qquad\qquad \frac{p}{m_0 v} = \frac{1}{\sqrt{1-v^2/c^2}}$$

$$\left(\frac{E}{m_0 c^2}\right)^2 = \frac{1}{\sqrt{1-v^2/c^2}} \qquad\qquad \left(\frac{p}{m_0 v}\right)^2 \frac{v^2}{c^2} = \frac{v^2/c^2}{\sqrt{1-v^2/c^2}}$$

$$\frac{E^2}{m_0^{\,2} c^4} = \frac{1}{\sqrt{1-v^2/c^2}} \quad (A.1) \qquad\qquad \frac{p^2}{m_0^{\,2} c^2} = \frac{v^2/c^2}{\sqrt{1-v^2/c^2}} \quad (A.2)$$

Subtracting term for term momentum Equation (A.2) from mass Equation (A.1), we obtain:

$$\frac{E^2}{m_0^{\,2} c^4} - \frac{p^2}{m_0^{\,2} c^2}\frac{c^2}{c^2} = \frac{1}{\sqrt{1-v^2/c^2}} - \frac{v^2/c^2}{\sqrt{1-v^2/c^2}} \qquad (A.3)$$

$$\frac{E^2}{m_0^{\,2} c^4} - \frac{p^2 c^2}{m_0^{\,2} c^4} = \frac{1}{\sqrt{1-v^2/c^2}} - \frac{v^2/c^2}{\sqrt{1-v^2/c^2}}$$

$$\frac{E^2 - p^2 c^2}{m_0^{\,2} c^4} = \frac{1-v^2/c^2}{1-v^2/c^2} = \frac{\gamma}{\gamma} \qquad (A.4)$$

$$\frac{E^2 - p^2 c^2}{m_0^{\,2} c^4} = \frac{\gamma^2}{\gamma^2}$$

$$\gamma^2\left(E^2 - p^2 c^2\right) = \gamma^2 m_0^{\,2} c^4$$

$$\gamma^2 E^2 - \gamma^2 p^2 c^2 = \left(mc^2\right)^2 \quad \text{where } \gamma m_o = m$$

$$\gamma^2 E^2 = (pc)^2 + \left(mc^2\right)^2 \quad \text{where } p = \gamma m_o v = mv \qquad (A.5)$$

And finally $E=\gamma E$ and we obtain Equation (2.41):

$$E^2 = (pc)^2 + (mc^2)^2 \qquad\qquad (2.41)$$

Considering step (A.4) during the derivation sequence, there is a strong temptation to simplify both occurrences of the γ Lorentz factor to 1 before proceeding, but this leads to the often encountered and erroneous non-relativistic version $E^2=(pc)^2+(m_oc^2)^2$, which is frequently given as being the ultimate representation of the Special Relativity theory, but which is indeed simply Newtonian, since such a simplification to 1 has for consequence that all occurrences of the γ factor disappear from the equation. Consequently, only the classical value of the ΔK momentum energy is then provided to the rest mass of the particle in motion, on top of leaving out the $\Delta m_m c^2$ magnetic energy component of the carrying energy that electromagnetically oscillates transversely and provides the velocity related relativistic mass increment which is transversely measurable.

The proper procedure is then to square the mutually reducible γ factor occurrences, so they can be reunited with the two occurrences of m_o as the development proceeds.

At face value, fusing the last occurrence of the squared γ factor with the squared energy (γ^2E^2) may seem to be problematic, but considering that this factor is a dimensionless quantity (See Section 3.5), it can be multiplied with the *energy* component without any adverse effect for the integrity of the equation and simply increases the total energy amount on the left side of the equation to the same relativistic value that it now has on the right side.

A note of caution must also be aired regarding the mathematically unsound and deeply anchored urban legend in the physics community that it suffices to set m to zero in the $(mc^2)^2$ term of Equation (2.41) to reduce the equation to $E=pc$, that would then supposedly provide the energy of a free moving photon.

This is ignoring the basic mathematical rule that if an element of an equation is set to zero in one of its terms, it has to also be set to zero in all other terms, including in term $(pc)^2$ in the present case, since steps (A.0) and (A.4) reveal that the momentum p symbol can only be defined in context as being equal to mv in the energy-momentum equation, and cannot be made equal to $\lambda v/c$ according to any logical derivation.

Moreover, the analysis carried out in this work reveals that to proceed in this manner is doubly erroneous because $p=mv$ provides only half the energy of the massive particle carrier-photon, that is, only its ΔK momentum energy half-quantum, while $p=\lambda v/c$ provides the total energy of a free moving photon, that is, its ΔK momentum energy half-quantum plus the $\Delta m_m c^2$ energy of its transversely oscillating electromagnetic half-quantum.

Consequently, proceeding to set m to zero only in the $(mc^2)^2$ mass term of the energy-momentum equation without setting it in context to zero in the $(pc)^2$ momentum term reveals a logical inconsistency tantamount to a level of mathematical illiteracy quite reminiscent of the logical inconsistency observed with defective Equation (7.1.2) found in Reference [7], that apparently drew no attention in the formal physics community, as analyzed in Section 1.7.2.

Mathematics is a language that must be learned to the same level of proficiency as for engineering applicability before fundamental physics issues are studied in depth, otherwise havoc such as that wrought by the Copenhagen interpretation is likely to affect the community again. Indeed, it can be observed that the scientists who most strongly influenced the evolution of theoretical physics, such as Gauss, Maxwell, Minkowski and Poincaré, all were in reality high-level mathematicians who coherently synthesized the equations that were established from confirmed experimental data obtained by hands-on experimentalists.

A.2. The Trispatial Energy-Momentum Equation

It should be noted that the traditional relativistic energy-momentum equation (2.41) is not used anywhere to make any calculations due to its complexity of resolution. The new trispatial energy-momentum Equation (1.50) however:

$$E_e = \Delta K + \Delta m_m c^2 + m_0 c^2 \tag{1.50}$$

is easy to use for any motion related energy calculation, since its two components ΔK and $\Delta m_m c^2$ are always equal by structure, and there is no need to use the γ factor to resolve it.

Contrary to relativistic equation (2.41), m_o appears in only one of its terms. Therefore, in case of Equation (1.50) it is effectively sufficient to set m_o to zero in the term $m_o c^2$ to reduce the equation to $E=\Delta K+\Delta m_m c^2$, which then effectively becomes Equation (2.13) of the particle's carrier-photon, which is also one of the standard equations to calculate the energy of an electromagnetic photon.

$$E = \Delta K + \Delta m_m c^2 \tag{2.13}$$

Knowing then the energy of every term of the equation, whether for Equation (1.50) or Equation (2.13), it becomes easy to calculate the velocity of the particle using one of the two equations (1.33):

$$v = c\frac{\sqrt{\lambda_c(4\lambda + \lambda_c)}}{(2\lambda + \lambda_c)} \quad \text{or} \quad v = c\frac{\sqrt{4EK + K^2}}{2E + K} \tag{1.33}$$

See also Section 3.5.1.

Appendix B

B.1. Maxwell's equations

Maxwell's Equations		
Atomic, Macroscopic and Astronomical orders of magnitude		Subatomic order of magnitude
Integral Form	Differential form	First Level Form
1 $\oint \mathbf{E} \cdot d\mathbf{S} = {}^{q}\!\big/\!{}_{\varepsilon_0} = \Phi_E$	$\nabla \cdot \mathbf{E} = \rho/\varepsilon_0$	$\mathbf{E}_\lambda = \dfrac{\pi e}{\varepsilon_0 \alpha^3 \lambda^2}$
2 $\oint \mathbf{E} \cdot d\mathbf{l} = -d\left(\int \mathbf{B} \cdot \hat{n} d\mathbf{S}\right)\!/dt = -d\Phi_B/dt$	$\nabla \times \mathbf{E} = -\partial \mathbf{B}/\partial t$	$v = \dfrac{\mathbf{E}_{\lambda_c} \times \Delta\mathbf{E}_\lambda}{\mathbf{B}_{\lambda_c} + \Delta\mathbf{B}_\lambda}$
3 $\oint \mathbf{B} \cdot d\mathbf{S} = 0$	$\nabla \cdot \mathbf{B} = 0$	$\mathbf{B}_\lambda = \dfrac{\mu_0 \pi e c}{\alpha^3 \lambda^2}$
4 $\oint \mathbf{B} \cdot d\mathbf{l} = \mu_0\left(\mathbf{i} + \varepsilon_0 d(\Phi_E)/dt\right)$	$\nabla \times \mathbf{B} = \mu_0\left(\mathbf{J} + \dfrac{\varepsilon_0 \partial \mathbf{E}}{\partial t}\right)$	$c = \dfrac{\mathbf{E}_\lambda}{\mathbf{B}_\lambda}$

B.2. Equations for the atomic, macroscopic and astronomical orders of magnitude

The set of equations known as Maxwell's equations were developed in reality by Gauss, Faraday and Ampere from physically carried out experiments. Maxwell's major contribution to science, after analysis of the observed fact that changing magnetic fields induce current in conducting wires, and that reciprocally, as previously discovered by Oersted, that current circulating in a wire induce a magnetic field around the wire, was his intuition that such mutual induction of electric and magnetic fields could occur in space without material supports such as magnets and electric wires.

This led him to relate this hypothesis to the enigma of light propagation, after Faraday informed him, as mentioned at the beginning of Section 1.1, that when he placed a glass plate between the poles of an electromagnet, the magnetic field caused the polarization plane of the light passing through the plate to rotate.

He then drew the conclusion that light had to be actual electromagnetic energy, and since the range of visible light frequencies was rather limited, that is, from about 405 THz for red light to about 790 THz for violet light, that this limited range had to be part of a potentially more complete spectrum with other frequencies, that would be invisible to us this time, and that would extend in both directions, that is, higher than the 790 THz of violet light and shorter than 405 THz for red light

His hypothesis in this regard was first confirmed 20 years later when Hertz confirmed the existence of radio frequencies. The rest is history, and his continuous wave theory of electromagnetic energy has proven completely successful in dealing with electromagnetic energy all the way from the atomic to the astronomical level of magnitude.

Maxwell's first equation is actually Gauss's equation for the electric field, which is a generalization of the Coulomb law, establishing a potential electric interaction field, by removing one charge from the Coulomb equation (See Subsection 1.7.1).

The second equation, derived from Faraday's law of induction means that a variation of a magnetic field is required for an electric field to be produced. In the context of the localized point-like fields of the present model, it can be interpreted without modification as meaning that any variation of the magnetic aspect of an electromagnetic event is mandatorily accompanied by a corresponding inverse variation of its electric aspect.

The third equation corresponds to Gauss's law for magnetism, that defines a potential magnetic interaction field as a counterpart to the potential electric field defined by the first equation, and implies that as much *magnetic* energy flows out of a given volume containing the source of the field as flows in, hence the resulting zero value.

The forth equation, derived from Ampere's law and named the Ampere-Maxwell equation, initially accounting for the observation that a magnetic field is produced by an electric current in a wire, which Maxwell then extended to the conclusion that a magnetic field is also produced by a changing electric field and reciprocally, even without material support, which constitutes Maxwell's greatest discovery.

B.3. Equations for the subatomic order of magnitude

The four first level electromagnetic equations for the subatomic order of magnitude were developed during the first wave of derivations after Paul Marmet's finding, and were published in 2007 in the *"International IFNA-ANS Journal"* at Kazan State University ([30], [8] Chapter 4).

The term *"first level"* refers to the fact that, contrary to Maxwell's equations as traditionally referred to in all reference works, and as presented above, the subatomic level equations are just one step removed from displaying the complete set of constants and variables, that can immediately be used to calculate a physical value, just like the Coulomb Equation (2.19). The analysis of the reason why the elaboration of such first level equations is required for progress to be made in fundamental physics was made in Section 27 of reference [25].

The first level Gauss electric equation was developed as Equation (40) in reference [30]. See Section 2.7 for an example of use:

$$\mathbf{E}_\lambda = \frac{\pi e}{\varepsilon_0 \alpha^3 \lambda^2} \tag{B.1}$$

as well as first level Gauss magnetic equation as Equation (34) in the same reference:

$$\mathbf{B}_\lambda = \frac{\mu_0 \pi e c}{\alpha^3 \lambda^2} \tag{B.2}$$

The first level electric composite *E* field equation required to calculate the velocity of a massive charged particle, which is in fact the completely resolved *E* field of the Lorentz equation $F = q(E + v \times B)$ was then resolved as Equation (58) in the same reference, and is here completely developed for convenience:

$$\mathbf{E} = \mathbf{E}_{\lambda_c} \times \Delta\mathbf{E}_\lambda = \frac{\pi e}{\varepsilon_0 \alpha^3} \frac{\left(\lambda^2 + \lambda_c^2\right)}{\lambda^2 \lambda_c^2} \frac{\sqrt{\lambda_c\left(4\lambda + \lambda_c\right)}}{\left(2\lambda + \lambda_c\right)} \tag{B.3}$$

The first level magnetic composite *B* field equation required to calculate the velocity of a massive charged particle, which is the completely resolved *B* field of the Lorentz equation, was resolved as Equation (49) in the same reference, and is completely developed here for convenience:

$$\mathbf{B} = \mathbf{B}_{\lambda_c} + \Delta\mathbf{B}_\lambda = \frac{\pi \mu_0 e c}{\alpha^3} \frac{\left(\lambda^2 + \lambda_c^2\right)}{\lambda^2 \lambda_c^2} \tag{B.4}$$

Equations (B.3) and (B.4) can then be used directly to calculate the velocity of a charged massive particle with traditional equation $v=E/B$. Similarly, Equations (B.1) and (B.2) can be used directly to calculate the velocity of any free moving photon with equation $c=E_\lambda/B_\lambda$.

Afterword

The aim of the present project was to explore the mechanical conversion processes involving electromagnetic energy and the very limited set of stable elementary electromagnetic particles that were confirmed to exist at the subatomic level of physical reality by means of non-destructive scattering, and that are the building blocks of all atoms, as well as the free moving electromagnetic photons that are emitted by these stable elementary particles as they stabilize within atomic structures and whose absorption causes them to temporarily or permanently modify their stationary equilibrium states, as well as the processes by which they stabilize into the observed hierarchy of these stable equilibrium states.

The next stage will involve the establishment of the various complex resonance wave functions involving the mix of fixed and varying beat frequencies that define the resonance volumes of each of these stable stationary electromagnetic equilibrium states according to the method described in Reference ([25] Section 27).

The method used to define the trispatial geometry and explore the subatomic level of physical reality to the level achieved in this project is analyzed and described in Reference [98]. As surprising as this may seem to most, anybody could have written this book, because Nature has equipped each of us with a personal sample of the most powerful correlator in existence, our neocortex. Reference [25] describes and explains why and how it can allow each of us to progressively cause each of our personal understanding of physical reality to progressively drift towards a state that would be as close as humanly possible to objective understanding of this physical reality, by the simple means of systematically confirming the validity of all elements chosen as a grounding set on which to draw each of our conclusions, because a multilayer neural network such as the neocortex is unable to provide an invalid conclusion from a set all elements of which are valid.

Reference [99] analyses and describes how each child can be guided into achieving as complete a mastery as possible of its personal correlator. Unfortunately, countless children are deprived of this guidance in my own community and in many others by formally trained pedagogues who remain unfamiliar with the accumulated knowledge and the understanding that was gained by the discoverers of the various aspects of our understanding ability. The outcome is that many tolerate, and in many cases even encourage, the prescription of mind-numbing drugs to control unruly behavior in children, one important cause of which is precisely this lack of appropriate formal guidance, as observed in a field study that was carried out in the elementary schools of a large city in my community, that can be accessed via Reference [100].

References

[1] Selleri, F. (1994) *Le grand débat de la théorie quantique.* Champs. Flammarion. France.

[2] Petkov, V. Editor. (2012) *Space and Time - Minkowski's Papers on Relativity.* Minkowski Institute Press. Montreal. Canada. https://www.amazon.com/Space-Time-Minkowskis-papers-relativity/dp/0987987143.

[3] Planck, M. (1931) *Positivismus und reale Aussenwelt.* Akademische Verlagsgeselschaft M. B. H., Leipzig. https://catalog.princeton.edu/catalog/2057791

[4] Einstein, A., Schrödinger, E., Pauli, W., Rosenfeld, L., Born, M., Joliot-Curie, I. & F., Heisenberg, W., Yukawa, H., et al. (1953) *Louis de Broglie, physicien et penseur.* 2e Éditions Albin Michel, Paris.

[5] Einstein A. (1910) *Le Principe de relativité et ses conséquences dans la physique moderne.* Traduit de l'allemand par E. Guillaume. Archives des sciences physiques et naturelle 29 (1910): 5-28; 125-144. http://www.minkowskiinstitute.org/mip/books/einstein2.html

[6] Pais, A. (2005) *Subtle is the Lord: The Science and the Life of Albert Einstein.* Oxford University Press. New York.

[7] Ciufolini I & Wheeler JA (1995). *Gravitation and Inertia,* Princeton University Press.

[8] Michaud A. (2017). *Electromagnetic Mechanics of Elementary Particles. 2nd Edition.* Scholar's Press. Saarbrücken, Germany. 2016. ISBN: 978-3-330-65345-0. https://www.morebooks.de/store/gb/book/electromagnetic-mechanics-of-elementary-particles/isbn/978-3-330-65345-0

[9] Michaud, A. (2020) *Electromagnetism according to Maxwell's Initial Interpretation.* Journal of Modern Physics, 11, 16-80. https://doi.org/10.4236/jmp.2020.111003. https://www.scirp.org/pdf/jmp_2020010915471797.pdf.

[10] Michaud, A. (2018) *The Hydrogen Atom Fundamental Resonance States.* Journal of Modern Physics,9,1052-1110.doi:10.4236/jmp.2018.95067.

https://file.scirp.org/pdf/JMP_2018042716061246.pdf.

[11] Michaud, A. (2017) *Gravitation, Quantum Mechanics and the Least Action Electromagnetic Equilibrium States*. J Astrophys Aerospace Technol 5: 152. doi:10.4172/2329-6542.1000152.

https://www.omicsonline.org/open-access/gravitation-quantum-mechanics-and-the-least-action-electromagneticequilibrium-states-2329-6542-1000152.pdf.

[12] Michaud, A. (2020) *Gravitation, Quantum Mechanics and the Least Action Electromagnetic Equilibrium States*. In: Amenosis Lopez, editor. Prime Archives in Space Research. Hyderabad, India: Vide Leaf. 2020.

https://videleaf.com/gravitation-quantum-mechanics-and-the-least-action-electromagnetic-equilibrium-states/

[13] Rousseau, P. (1959) *La Lumière*. Presses Universitaires de France, Collection "Que sais-je?". France.

[14] Michaud, A. (2013) *Deriving Eps_0 and Mu_0 from First Principles and Defining the Fundamental Electromagnetic Equations Set*. International Journal of Engineering Research and Development e-ISSN: 278-067X, p-ISSN: 2278-800X, Volume 7, Issue 4 (May 2013), PP. 32-39.

http://ijerd.com/paper/vol7-issue4/G0704032039.pdf.

[15] Michaud, A. (2016) *On De Broglie's Double-particle Photon Hypothesis*. J Phys Math 7: 153. doi:10.4172/2090-0902.1000153,

https://www.omicsonline.org/open-access/on-de-broglies-doubleparticle-photon-hypothesis-2090-0902-1000153.pdf.

[16] Cornille, P. (2003) *Advanced Electromagnetism and Vacuum Physics*. World Scientific Publishing, Singapore.

[17] Sears F., Zemansky M., Young H. (1984) *University Physics*, 6th Edition, Addison Wesley.

[18] Eisberg, R., and Resnick, R. (1985) *Quantum Physics of Atoms, Molecules, Solids, Nuclei, and Particles*. 2nd Edition, John Wiley & Sons, New York.

[19] Griffiths, D.J. (1999) *Introduction to Electrodynamics*. Prentice Hall, USA.

[20] Jackson, J.D. (1999) *Classical Electrodynamics*. John Wiley & Sons. USA.

[21] Breidenbach M. et al. (1969) *Observed Behavior of Highly Inelastic Electron-Proton Scattering*, Phys.Rev.Let.,Vol.23,No.16,935-939.
https://journals.aps.org/prl/abstract/10.1103/PhysRevLett.23.935.

[22] Ohanian, H.C., Ruffini, R. (1994) *Gravitation and Spacetime*, Second Edition, W.W. Norton. P. 194.

[23] Anderson, C.D. (1933) *The Positive Electron*. Phys.Rev.43,491.
https://journals.aps.org/pr/pdf/10.1103/PhysRev.43.491.

[24] McDonald, K., et al. (1997) Positron Production in Multiphoton Light-by-Light Scattering, Phys.Rev.Lett.79,1626.
http://www.slac.stanford.edu/exp/e144/.
http://journals.aps.org/prl/abstract/10.1103/PhysRevLett.79.1626.

[25] Michaud, A. (2019) *The Mechanics of Conceptual Thinking*. Creative Education, 10, 353-406.
https://doi.org/10.4236/ce.2019.102028.
http://www.scirp.org/pdf/CE_2019022016190620.pdf.

[26] Feynman R.P., Leighton R.B and Sands M. (1964) *The Feynman Lectures on Physics*. Addison-Wesley, Vol. II, p. 28-1.

[27] De Broglie, L. (1993) *La physique nouvelle et les quanta*, Flammarion, France 1937, 2nd Edition 1993, with new 1973 Preface by Louis de Broglie. ISBN: 2-08-081170-3.

[28] Michaud, A. (2000) *On an Expanded Maxwellian Geometry of Space*. Proceeding of Congress-2000. "Fundamental Problems of Natural Sciences and Engineering". St Peterburg State University. Russia. Volume 1. pp. 291-310.

[29] Marmet, P. (2003) *Fundamental Nature of Relativistic Mass and Magnetic Fields*. International IFNA-ANS Journal, 9. 64-76. Kazan State University, Kazan, Russia.
http://www.newtonphysics.on.ca/magnetic/index.html.

[30] Michaud, A. (2007) *Field Equations for Localized Individual Photons and Relativistic Field Equations for Localized Moving Massive Particles*, International IFNA-ANS Journal, No. 2 (28), Vol. 13, 2007, p. 123-140, Kazan State University, Kazan, Russia. https://www.gsjournal.net/Science-Journals/Research%20Papers-Relativity%20Theory/Download/2257.

[31] Michaud, A. (2013) *The Mechanics of Electron-Positron Pair Creation in the 3-Spaces Model*. International Journal of Engineering Research and

Development e-ISSN: 2278-067X, p-ISSN: 2278-800X, Volume 6, Issue 10 (April 2013), PP. 36-49.

http://ijerd.com/paper/vol6-issue10/F06103649.pdf.

[32] Michaud, A. (2013) *The Mechanics of Neutron and Proton Creation in the 3-Spaces Model*. International Journal of Engineering Research and Development e-ISSN: 2278-067X, p-ISSN : 2278-800X, Volume 7, Issue 9 (July 2013), PP.29-53.

http://www.ijerd.com/paper/vol7-issue9/E0709029053.pdf.

[33] Michaud, A. (2013) *The Mechanics of Neutrinos Creation in the 3-Spaces Model*. International Journal of Engineering Research and Development. e-ISSN: 2278-067X, p-ISSN: 2278-800X, Volume 7, Issue 7 (June 2013), PP. 01-08. http://www.ijerd.com/paper/vol7-issue7/A07070108.pdf.

[34] Michaud, A. (2017). *The Last Challenge of Modern Physics*. J Phys Math 8: 217. doi: 10.4172/2090-0902.1000217.

https://www.omicsonline.org/open-access/the-last-challenge-of-modern-physics-2090-0902-1000217.pdf.

[35] Bartels, J., Haidt, D., Zichichi, A. Editors (2000) *The European Physical Journal C - Particles and fields*. Springer, Germany.

[36] Kaufmann, W. (1903) *Über die "Elektromagnetische Masse" der Elektronen*, Kgl. Gesellschaft der Wissenschaften Nachrichten, Mathem.-Phys. Klasse, pp. 91-103.
http://gdz.sub.unigoettingen.de/dms/load/img/?PPN=PPN252457811_1903&DMDID=DMDLOG_0025.

[37] Lorentz, H.A. (1904) *Electromagnetic phenomena in a system moving with any velocity smaller than that of light*, in: KNAW, Proceedings, 6, 1903-1904, Amsterdam, 1904, pp. 809-831.

https://en.wikisource.org/wiki/Electromagnetic_phenomena.

[38] Einstein, A. (1934) *Comment je vois le monde*, Flammarion, France, 1958.

[39] Abraham, M. (1902) *Dynamik des Elektrons*, Nachrichten von der Gesellschaft der Wissenschaften zu Göttingen, Mathematisch-Physikalische Klasse,1902,S.20.
http://gdz.sub.unigoettingen.de/dms/load/img/?PPN=PPN252457811_1902&DMDID=DMDLOG_0009.

[40] Poincaré, H. (1902) *La science et l'hypothèse*, France, Flammarion 1902, 1995 Edition.

[41] Planck, M. (1906) *Das Prinzip der Relativität und die Grundgleichungen der Mechanik.* Verhandlungen Deutsche Physikalische Gesellschaft. 8, pp. 136–141. (Vorgetragen in der Sitzung vom 23. März 1906.).
https://archive.org/details/verhandlungende00goog/page/n179.

[42] Michaud, A. (2013) *From Classical to Relativistic Mechanics via Maxwell,* International Journal of Engineering Research and Development, e-ISSN: 2278-067X, p-ISSN: 2278-800X. Volume 6, Issue 4. pp. 01-10.
http://www.gsjournal.net/Science-Journals/Essays/View/3197.

[43] Michaud, A. (2016) *On Adiabatic Processes at the Elementary Particle Level.* J Phys Math 7: 177. doi: 10.4172/2090-0902. 1000177.
https://www.omicsonline.org/open-access/on-adiabatic-processes-at-the-elementary-particle-level-2090-0902-1000177.pdf.

[44] Michaud, A. (2013) *Unifying All Classical Force Equations,* International Journal of Engineering Research and Development, e-ISSN: 2278-067X, p-ISSN: 2278-800X, Volume 6, Issue 6 (March 2013), PP. 27-34.
http://www.ijerd.com/paper/vol6-issue6/F06062734.pdf.

[45] Michaud, A. (2013) *Inside Planets and Stars Masses.* International Journal of Engineering Research and Development e-ISSN: 2278-067X, p-ISSN: 2278-800X, Volume 8, Issue 1 (July 2013), PP. 10-33.
http://ijerd.com/paper/vol8-issue1/B08011033.pdf.

[46] Anderson, J.D., Laing, A., Lau, E.L., Liu, A.S., Nieto, M.M. et al. (1998) Indications from Pioneer 10/11, Galileo, and Ulysses Data, of an Apparent Anomaleous, Weak, Long-Range Acceleration, gr-qc/9808081, v2, 1 Oct 1998.
http://arxiv.org/pdf/gr-qc/9808081v2.pdf.

[47] Nieto, M.M., Goldman, T., Anderson, J.D., Lau, E.L., Perez-Mercader, J. (1994) *Theoretical Motivation for Gravitation Experiments on Ultra low Energy Antiprotons and Antihydrogen,* hep-ph/9412234, 5 Dec 1994.
http://arxiv.org/pdf/hep-ph/9412234.pdf.

[48] Anderson, J.D., Campbell, J.K, Nieto, M.M. (2006) *The energy transfer process in planetary flybys,* astro-ph/0608087v2, 2 Nov 2006.
http://arxiv.org/pdf/astro-ph/0608087.pdf.

[49] Michaud, A. (2013) *On The Magnetostatic Inverse Cube Law and Magnetic Monopoles.* International Journal of Engineering Research and

Development e-ISSN: 2278-067X, p-ISSN: 2278-800X. Volume 7, Issue 5. pp. 50-66.

http://www.ijerd.com/paper/vol7-issue5/H0705050066.pdf.

[50]   National Institute of Standards and Technology, (NIST).

https://www.physics.nist.gov/cgi-bin/cuu/Value?h|search_for=universal_in!.

[51]   Lide, D.R., Editor-in-chief. (2003) *CRC Handbook of Chemistry and Physics*. 84thEdition 2003-2004, CRC Press, New York. 2003.

[52]   Michaud, A. (2013) *On the Einstein-de Haas and Barnett Effects*, International Journal of Engineering Research and Development. e-ISSN: 2278-067X, p-ISSN: 2278-800X, Volume 6, Issue 12, pp. 07-11.

http://ijerd.com/paper/vol6-issue12/B06120711.pdf.

[53]   Michaud, A. (2013) *The Expanded Maxwellian Space Geometry and the Photon Fundamental LC Equation*. International Journal of Engineering Research and Development, e-ISSN: 2278-067X, p-ISSN: 2278-800X. Volume 6, Issue 8, pp. 31-45.

http://ijerd.com/paper/vol6-issue8/G06083145.pdf.

[54]   Kühne, R.W. (1998) Remark on "Indication, from Pioneer 10/11, Galileo, and Ulysses Data, of an Apparent Anomalous, Weak, Long-Range Acceleration". arXiv:gr-qc/9809075v1 28 Sep 1998.

https://arxiv.org/pdf/gr-qc/9809075.pdf.

[55]   Hafele, J.C., and Keating, R.E. (1972) *Around-the-World Atomic Clocks: Predicted Relativistic Time Gains*. Science, New Series, Vol. 177, No. 4044, pp. 166-168. DOI: 10.1126/science.177.4044.166. http://www.personal.psu.edu/rq9/HOW/Atomic_Clocks_Experiment.pdf.

[56]   Michaud, A. (2016) *On the Birth of the Universe and the Time Dimension in the 3-Spaces Model*. American Journal of Modern Physics. Special Issue: Insufficiency of Big Bang Cosmology. Vol. 5, No . 4-1, 2016, pp. 44-52. doi: 10.11648/j.ajmp.s.2016050401.17.

http://article.sciencepublishinggroup.com/pdf/10.11648.j.ajmp.s.2016050401.17.pdf.

[57]   Resnick, R., & Halliday, D. (1967) *Physics*. John Wyley & Sons, New York.

[58]   De Broglie, L. (1923) *Ondes et Quanta*. Comptes rendus T.177 (1923) 507-510.

http://www.academie-sciences.fr/pdf/dossiers/Broglie/Broglie_pdf/CR1923_p507.pdf.

[59]   Kaku, M. (1993) *Quantum Field Theory*. Oxford University Press. New York.

[60]   Michaud, A. (2013) *On the Electron Magnetic Moment Anomaly*, International Journal of Engineering Research and Development. e-ISSN: 2278-067X, p-ISSN: 2278-800X. Volume 7, Issue 3, PP. 21-25.

http://ijerd.com/paper/vol7-issue3/E0703021025.pdf.

[61]   Michaud, A. (2013) *The Corona Effect*. International Journal of Engineering Research and Development e-ISSN: 2278-067X, p-ISSN: 2278-800X, Volume 7, Issue 11(July2013), PP. 01-09.

http://www.ijerd.com/paper/vol7-issue11/A07110109.pdf.

[62]   Lowrie, W. (2007) *Fundamentals of Geophysics*, Second Edition, Cambridge University Press.

[63]   Auger, A., Ouellet, C. (1998) *Vibrations, ondes, optique et physique moderne*. 2e Édition. Le Griffon d'argile. Quebec. Canada.

http://collegialuniversitaire.groupemodulo.com/2252-vibrations-ondes-optique-et-physique-moderne-2e-edition-produit.html.

[64]   Kotler S., Akerman N., Navon N., Glickman Y., Ozeri R. (2014) *Measurement of the magnetic interaction between two bound electrons of two separate ions*. Nature magazine. doi:10.1038/nature13403. Macmillan Publishers Ltd. Vol. 510, pp. 376-380. http://www.nature.com/articles/nature13403.epdf?referrer_access_token=yoC6RXrPyxwvQviChYrG0tRgN0jAjWel9jnR3ZoTv0PdPJ4geER1fKVR1YXH8GThqECstdb6e48mZm0qQo2OMX_XYURkzBSUZCrxM8VipvnG8FofxB39P4lc-1UIKEO1.

[65]   De Broglie, L. (1924) *Sur la définition générale de la correspondance entre onde et mouvement*, Comptes rendus de l'Académie des Sciences. (Paris) 179, 39.

[66]   De Broglie, L. (1924) *Sur un théorème de Bohr*, C. R. Acad. Sci. (Paris) 179, 676, Comptes rendus de l'Académie des Sciences. (Paris) 179, 39.

[67]   Schrödinger, E. (1952) *Are there quantum jumps?* Brit. J. Philos. Sci. 3 109,233.

https://philpapers.org/rec/SCHATQ-3

[68] Golovko, V.A. (2008) Electromagnetic radiation and resonance phenomena in quantum mechanics. arXiv:0810.3773v2.

https://arxiv.org/abs/0810.3773

[69] Schrödinger, E. (1930) *Über die kräftefreie Bewegung in der relativistischen Quantenmechanik*, Sitzungsberichte Akad. Berlin 1930, 418-428.

[70] Schwinger, J. (1948) On Quantum-electrodynamics and the Magnetic Moment of the Electron. Phys. Rev. 73, 416-417.

[71] Haskell, R.E. (2003) *Special Relativity and Maxwell's Equations*, Computer Science3 and Engineering Department, Oakland University, Rochester, Mi 48309.

http://www.cse.secs.oakland.edu/haskell/Special%20Relativity%20and%20Maxwells%20Equations.pdf

[72] Ernst, A. and Hsu, J.P. (2001) *First Proposal of the Universal Speed of Light by Voigt in 1887*, Chinese Journal of Physics, Vol. 39, No. 3.

http://adsabs.harvard.edu/cgi-bin/nph-data_query?bibcode=2001ChJPh..39..211E&link_type=ARTICLE&db_key=PHY&high=

[73] Particle Data Group. The European Physical Journal - Review of Particle Physics, Volume 15 – Number 10-4.2000.

[74] Cauchois Y. (1952). *Atomes, Spectres, Matière*. Éditions Albin Michel, Paris, 1952.

[75] Poincaré H. (1905). *La valeur de la science*, France, Flammarion 1994 Edition.

[76] Poincaré, M.H. (1905) *Sur la dynamique de l'électron*. Comptes rendus de l'Académie française. 1905/01 (T140)-1905/06, pp 1504-1508.

[77] Poincaré, M.H. (1906) *Sur la dynamique de l'électron*. Rendiconti del circolo matematico di Palermo **21,** 129–175. https://doi.org/10.1007/BF03013466.

https://fr.wikisource.org/wiki/Sur_la_dynamique_de_l%E2%80%99%C3%A9lectron

[78] Blackett P.M.S. & Occhialini G. (1933). *Some photographs of the tracks of penetrating radiation*, Proceedings of the Royal Society, 139, 699-724.

[79] Anderson J.D. et al. (2005). Study of the anomalous acceleration of Pioneer 10 and 11, gr-qc/0104064.

https://arxiv.org/abs/gr-qc/0104064

[80] Keith J.C. (1963). *Gravitational Radiation and Aberrated Cenripetal force Reactions in Relativity theory. Part 2.* Retarded cohesive Forces. Revista Mexicana de Fisica. Vol.XII,1: (7 Marzo de 1963).

[81] Fremerey J.K. (1973). Significant Deviation of Rotational Decay from Theory at a Reliability in the 10^{-12} sec^{-1} Range. Phys. Rev. Lett., v. 30, no. 16, pp. 753-757.

https://journals.aps.org/prl/abstract/10.1103/PhysRevLett.30.753

[82] Blewett J.P. (1946). Radiation Losses in the Induction Electron Accelerator, Phys. Rev. 69, 87.

https://journals.aps.org/pr/abstract/10.1103/PhysRev.69.87

[83] Turner, S. Editor. (1994) *CERN Accelerator School — Fifth General Accelerator Physics Course.* Proceedings. University of Jyväskylä. Finland.

https://cds.cern.ch/record/235242/files/CERN-94-01-V1.pdf.

[84] Storti, R. (2011) Quinta Essentia. *A Practical Guide to Space-Time Engineering.* Delta Group Engineering. Australia.

[85] Giancoli, D.C., (2008) *Physics for Scientists & Engineers.* Pearson Prentice Hall, USA.

[86] Çengel, Y.A., & Boles, M.A., (2002) *Thermodynamics - An Engineering Approach.* McGraw Hill, USA.

[87] Meriam, J.L., & Kraige, L.G., (2003) *Engineering Mechanics Dynamics.* John Wiley and Sons. USA.

[88] Rao, S.S., (2005) *Mechanical Vibrations.* Pearson Prentice Hall, Singapore.

[89] Rao, N.N. (2000) *Elements of Engineering Electromagnetics.* 5th Edition. Prentice Hall. Upper Saddle River, New Jersey.

[90] Hibbeler, R.C., (2005) *Mechanics of Materials.* Pearson Prentice Hall, USA.

[91] De Broglie L. (1934). *L'équation d'ondes du photon,* C. R. Acad. Sci., **199**, p. 445-448.

[92] De Broglie L. and Winter M.J. (1934). *Sur le spin du photon*, C. R. Acad. Sci., **199**, p. 813-816.

[93] De Broglie L. (1936). La théorie du photon et la mécanique ondulatoire relativiste des systèmes, C. R. Acad. Sci., 203, p. 473-477.

[94] De Broglie L. (1937). *La quantification des champs en théorie du photon*, C. R. Acad. Sci., **205**, p. 345-349.

[95] Markoulakis, E., Rigakis, I., Chatzakis, J., Konstantaras, A., Antonidakis, E. (2018) *Real time visualization of dynamic magnetic fields with a nanomagnetic ferrolens*, J. Magn. Magn. Mater. 451 (2018) 741-748. doi:10.1016/j.jmmm.2017.12.023.

https://www.sciencedirect.com/science/article/abs/pii/S0304885317319194?via%3Dihub.

[96] Soosaleon A. (2017). *Gravity Induced Resonant Emission.* arXiv:1704.07225v1 [physics.plasm-ph+2] 4 Apr 2017.
https://arxiv.org/pdf/1704.07225.pdf

[97] Born M. & Fock V. (1928). *Beweis des Adiabatensatzes*. In: Zeitschrift für Physik. Band 51, Nr. 3-4, März 1928, S. 165–180, doi:10.1007/BF01343193.
https://link.springer.com/article/10.1007%2FBF01343193

[98] Michaud A (2017) On the Relation between the Comprehension Ability and the Neocortex Verbal Areas. J Biom Biostat 8: 331. doi:10.4172/2155-6180.1000331
https://www.hilarispublisher.com/open-access/on-the-relation-between-the-comprehension-ability-and-the-neocortexverbal-areas-2155-6180-1000331.pdf.

[99] Michaud A (2016) *Intelligence and Early Mastery of the Reading Skill*. J Biom Biostat 7: 327. doi: 10.4172/2155-6180.10003.
https://www.hilarispublisher.com/open-access/intelligence-and-early-mastery-of-the-reading-skill-2155-6180-1000327.pdf.

[100] Michaud A (2016) *Critical Analysis of a Field Research Report on ADD and ADHD*. Int J Swarm Intel Evol Comput 5: 142. doi: 10.4172/2090-4908.1000142.
https://www.longdom.org/open-access/critical-analysis-of-a-field-research-report-on-add-and-adhd-2090-4908-1000142.pdf.

9 789975 323833